# THE

## Orchards, Berry Patches and Gardens
# COOKBOOK

by
Bruce Carlson

Heart Huggin' 'N Tummy Warmin'
recipes from

*HEARTS & TUMMIES*
*COOKBOOK COMPANY*
-*a Dinky Division of Quixote Press*

(800) 571-BOOK
(319) 372-7480

© 1997 B. Carlson

All rights reserved. No part of this book may be reproduced or transmitted in any form or by any means, electronic or mechanical, including photocopying, recording or by any informational storage or retrieval system, except by a reviewer who may quote brief passages in a review to be printed in a magazine or newspaper - without permission in writing from the publisher.

* * * * * * * * * * * *

Although the author has exhaustively researched all sources to ensure the accuracy and completeness of the information contained in this book, he assumes no responsibility for errors, inaccuracies, ommissions, or any inconsistency herein. Any slights of people or organizations are unintentional. Readers should consult an attorney or accountant for specific applications to their individual publishing ventures.

*HEARTS 'N TUMMIES COOKBOOK CO.*
615 Avenue H
Fort Madison, Iowa 52627
1-800-571-BOOK

# INTRODUCTION

Sorry, but I got funner things to do than write a silly introduction.

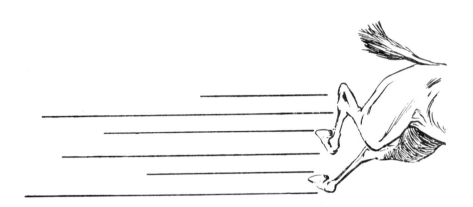

# FOREWORD

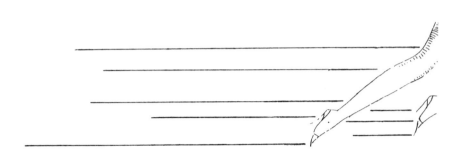

# TABLE OF CONTENTS

Fruit .................................................................................. 11
Vegetables ........................................................................ 65
Honey ............................................................................. 155
Nuts ................................................................................ 161
Order Form ..................................................................... 187

## PREFACE

It's nothing short of magic .... to turn soil, water, sunlight and work into delicious and beautiful bounty of the earth. I don't care what the chemists and the economists call it ... I call it beautiful.

And, that's fine. But, there's a hitch. After all that happens, what are you going to do with it? How are you going to turn all those fruits, veggies, honey and nuts into something for the table?

Well, that's what this cookbook is all about. It's a road map to take a person from a van full of produce to some of the yummiest eating on this side of anywhere.

> Prof. Phil Hey
> Briar Cliff College
> Sioux City, Iowa

# FRUIT

| | |
|---|---|
| Apple and Yam Casserole . . . . . . . . . . . . . . . 23 | Concord Grape Pie . . . . . . . . . . . . . . . . . . . . . 63 |
| Apple Bread . . . . . . . . . . . . . . . . . . . . . . . . . 29 | Crab Apple Pickles . . . . . . . . . . . . . . . . . . . . . 24 |
| Apple Butter . . . . . . . . . . . . . . . . . . . . . . . . . 15 | Cranberry Gelatin Salad . . . . . . . . . . . . . . . . . 60 |
| Apple Butter . . . . . . . . . . . . . . . . . . . . . . . . . 24 | Cranberry Salad . . . . . . . . . . . . . . . . . . . . . . . 63 |
| Apple Cherry Drink . . . . . . . . . . . . . . . . . . . 61 | Cranberry Salad . . . . . . . . . . . . . . . . . . . . . . . 60 |
| Apple Cinnamon Pie . . . . . . . . . . . . . . . . . . 31 | Cream Fruit Pops . . . . . . . . . . . . . . . . . . . . . . 57 |
| Apple Crisp . . . . . . . . . . . . . . . . . . . . . . . . . . 18 | Creamy Peach Pie . . . . . . . . . . . . . . . . . . . . . 39 |
| Apple Crisp . . . . . . . . . . . . . . . . . . . . . . . . . . 27 | Dip for Fruit Slices . . . . . . . . . . . . . . . . . . . . 33 |
| Apple Crisp . . . . . . . . . . . . . . . . . . . . . . . . . . 28 | Dutch Apple Pie . . . . . . . . . . . . . . . . . . . . . . 14 |
| Apple Crisp . . . . . . . . . . . . . . . . . . . . . . . . . . 29 | French Apple Cream Pie . . . . . . . . . . . . . . . . 13 |
| Apple Crisp Pudding . . . . . . . . . . . . . . . . . . . 28 | Fresh Blueberry Glaze Pie . . . . . . . . . . . . . . . 51 |
| Apple Crumb Pie . . . . . . . . . . . . . . . . . . . . . 13 | Fresh Gooseberry Pie . . . . . . . . . . . . . . . . . . 64 |
| Apple Crunch . . . . . . . . . . . . . . . . . . . . . . . . 21 | Fresh Peach Pie . . . . . . . . . . . . . . . . . . . . . . . 36 |
| Apple Crunch . . . . . . . . . . . . . . . . . . . . . . . . 27 | Fresh Peach Pie . . . . . . . . . . . . . . . . . . . . . . . 42 |
| Apple Pudding Bar . . . . . . . . . . . . . . . . . . . . 26 | Fresh Strawberry Pie . . . . . . . . . . . . . . . . . . . 43 |
| Apple Fruit Salad . . . . . . . . . . . . . . . . . . . . . 23 | Fried Apples . . . . . . . . . . . . . . . . . . . . . . . . . 30 |
| Apple Pudding . . . . . . . . . . . . . . . . . . . . . . . 16 | Fried Peaches . . . . . . . . . . . . . . . . . . . . . . . . 37 |
| Apple Rum . . . . . . . . . . . . . . . . . . . . . . . . . . 21 | Fruit Cobbler . . . . . . . . . . . . . . . . . . . . . . . . 61 |
| Apple Salad Dressing . . . . . . . . . . . . . . . . . . 22 | Fruit Cobbler . . . . . . . . . . . . . . . . . . . . . . . . 62 |
| Apple Slices . . . . . . . . . . . . . . . . . . . . . . . . . 22 | Fruit Dip . . . . . . . . . . . . . . . . . . . . . . . . . . . . 35 |
| Applesace Cake . . . . . . . . . . . . . . . . . . . . . . 20 | Fruit Dip . . . . . . . . . . . . . . . . . . . . . . . . . . . . 39 |
| Applesauce Cake . . . . . . . . . . . . . . . . . . . . . 26 | Fruit Pizza . . . . . . . . . . . . . . . . . . . . . . . . . . . 41 |
| Applesauce/Cinnamon Salad . . . . . . . . . . . . 54 | Fruit Pizza . . . . . . . . . . . . . . . . . . . . . . . . . . . 45 |
| Baked Peach Pudding . . . . . . . . . . . . . . . . . . 42 | Fruit Soup . . . . . . . . . . . . . . . . . . . . . . . . . . . 44 |
| Baked Pears . . . . . . . . . . . . . . . . . . . . . . . . . 34 | Fruited Dessert . . . . . . . . . . . . . . . . . . . . . . . 46 |
| Berried Delight . . . . . . . . . . . . . . . . . . . . . . 55 | Ginger Pears . . . . . . . . . . . . . . . . . . . . . . . . . 32 |
| Berry Delight . . . . . . . . . . . . . . . . . . . . . . . . 44 | Glaze Strawberry Pie . . . . . . . . . . . . . . . . . . 54 |
| Blackberry Pie . . . . . . . . . . . . . . . . . . . . . . . 64 | Glazed Blueberry Pie . . . . . . . . . . . . . . . . . . 51 |
| Black Raspberry Jelly . . . . . . . . . . . . . . . . . . 50 | Golden Peach Pie . . . . . . . . . . . . . . . . . . . . . 39 |
| Blueberry Buckle . . . . . . . . . . . . . . . . . . . . . 56 | Gotcha Peach Punch . . . . . . . . . . . . . . . . . . 41 |
| Blueberry Muffins . . . . . . . . . . . . . . . . . . . . 56 | Great Peach Cream Pie . . . . . . . . . . . . . . . . 42 |
| Brandied Apple Bars . . . . . . . . . . . . . . . . . . 17 | Hot Spiced Apple Cider . . . . . . . . . . . . . . . . 20 |
| Fancy Baked Apples . . . . . . . . . . . . . . . . . . 14 | Iced Apple Brownies . . . . . . . . . . . . . . . . . . 25 |
| Caramel Apples . . . . . . . . . . . . . . . . . . . . . . 16 | Mashed Potatoes with Apple . . . . . . . . . . . . 25 |
| Cherry Coffee Cake . . . . . . . . . . . . . . . . . . . 62 | Michigan Blueberry Buckle . . . . . . . . . . . . . 57 |
| Cherry Pudding . . . . . . . . . . . . . . . . . . . . . . 53 | Mom's Strawberry Stuff . . . . . . . . . . . . . . . . 43 |
| Cherry Pudding Cobbler . . . . . . . . . . . . . . . 59 | More Fresh Peach Pie . . . . . . . . . . . . . . . . . 36 |
| Chewy Fruit Leather . . . . . . . . . . . . . . . . . . 58 | More Strawberry/Rhubarb Pie . . . . . . . . . . . 48 |
| Chicken Fruit /n Nut Salad . . . . . . . . . . . . . 15 | Orange Juice Squeezin's . . . . . . . . . . . . . . . . 59 |
| Concord Grape Pie . . . . . . . . . . . . . . . . . . . 52 | Pastry Apple Bars . . . . . . . . . . . . . . . . . . . . . 19 |
| | Peach Apple Conserve . . . . . . . . . . . . . . . . 40 |

| | |
|---|---|
| Peach Cheese Dessert | 37 |
| Peach Cobbler | 35 |
| Peach Upside Down Cake | 40 |
| Pear Bread | 32 |
| Rainbow Roll | 64 |
| Raspberry Cobbler | 53 |
| Raw Apple Bars | 18 |
| Red Apple Cole Slaw | 30 |
| Red Raspberry Pie | 50 |
| Sour Cream Apple Squares | 17 |
| Spicy Apple Cookies | 19 |
| Still More Fresh Peach Pie | 38 |
| Strawberry Cream Dip | 47 |
| Strawberry Delight | 47 |
| Strawberry Graham Dessert | 49 |
| Strawberry Ice Cream Delight | 46 |
| Strawberry/Rhubarb Pie | 48 |
| Strawberry Shortcake | 49 |

## APPLE CRUMB PIE

1 unbaked pie shell
6-8 baking apples

TOPPING:
½ C. sugar
¼ tsp. salt

½ C. sugar
½ tsp. cinnamon

¾ C. flour
⅓ C. butter or margarine

Peel and slice apples and put in crust in layers, each time sprinkling with some sugar and cinnamon. Mix topping together, cutting in butter. Scatter topping over apples and press down lightly. Bake in 375° oven until done, about 1 hour. Serve with ice cream or whipped cream.

## FRENCH APPLE CREAM PIE

1 unbaked crust
4 C. sliced apples
1½ C. sugar
¾ C. milk

1 beaten egg
3 T. flour
1 tsp. vanilla
¼ tsp. nutmeg

TOPPING:
3 T. brown sugar
2 T. flour

1 T. soft butter

Mix milk, sugar, egg, flour, vanilla and nutmeg. Place apples in unbaked shell. Pour cream mixture over apples. Mix topping ingredients until crumbly and sprinkle on top of pie. Bake at 400° for 40-45 minutes.

## DUTCH APPLE PIE

1 unbaked pie shell
8 C. sliced apples
1 C. sugar
1⅓ tsp. cinnamon

1 C. flour
½ C. brown sugar
½ C. butter

Mix apples, sugar, and cinnamon. Place in pie shell. Mix flour, brown sugar, and butter until crumbly. Sprinkle on top of apples. Bake at 375° for 35-40 minutes.

## Fancy Baked Apples

4 apples
⅓ C. chopped almonds
¼ C. raisins

4 T. brown sugar
4 T. chopped nuts

Core but do not cut through the bottom of the apples. Combine the remaining ingredients and fill apples. Add ¼-inch water. Cover and bake at 350° for 30 minutes or until apples are done.

## CHICKEN FRUIT 'N NUT SALAD

3 C. torn lettuce leaves  
5 oz. Swanson premium chunk chicken (drained)  
1 apple (cored and sliced)  
½ C. seedless red grapes (sliced in half)  
¼ C. Ranch salad dressing (or more)  
Chopped walnuts

Mix first 4 ingredients. Toss with dressing lightly. Garnish with nuts.

## Apple Butter

2 qts. cooked apple juice  
2 tsp. cinnamon  
4 C. sugar  
¼ tsp. cloves

Measure pulp, add sugar and spices. Cook until flavors are well blended, about 15 minutes, stirring constantly to prevent sticking. Cook until mixture thickens. If too thick, add a small amount of water for desired consistency. Pour hot into hot ball jars, leaving ¼-inch head space. Adjust flats and bands. Process pints and quarts 10 minutes in canner of boiling water.

## Caramel Apples

45 caramels (unwrapped)
2 T. hot water
5 medium apples

5 wooden sticks
½ C. chopped peanuts

Place caramels and water in a 1-quart casserole. Microwave on high for 2 to 3 minutes until melted. Stir to make caramels smooth. Insert sticks in apples and dip in mixture. Now dip in peanuts, if desired. Place apples, stick side up on buttered wax paper. If mixture hardens, while coating apples, melt again for 25 to 35 seconds.

## Apple Pudding

1 egg
¾ C. sugar
2 T. flour
1¼ tsp. baking powder

1/8 tsp. salt
1 tsp. vanilla
½ C. chopped nuts
½ C. chopped apple

Beat egg until light; gradually beat in sugar. Stir in flour mixed with baking powder and salt. Add vanilla, nuts, and apple. Mix well and bake in buttered 8-inch pie pan for 35 minutes at 350°. Cut in wedges and serve warm with cream or ice cream. The pudding will be crumbly.

## SOUR CREAM APPLE SQUARES

2 C. flour
2 C. packed brown sugar
1-2 tsp. cinnamon
½ C. soft oleo
1 C. chopped nuts
2 C. peeled, finely chopped apples

1 tsp. soda
½ tsp. salt
1 C. dairy sour cream
1 tsp. vanilla
1 egg

In large bowl combine first 3 ingredients. Blend at low speed until crumbly. Stir in nuts. Press 2¾ C. crumb mixture in ungreased 9x13-inch pan. Blend rest of crumbs and remaining ingredients. Stir in apples. Spoon evenly over base. Bake 25-35 minutes at 350°. (Use toothpick test.)

## BRANDIED APPLE BARS

1⅓ C. flour
1 C. quick oatmeal
½ C. packed brown sugar
¾ C. butter
½ C. chopped nuts

8 oz. cream cheese (soft)
7 oz. jar marshmallow creme
2 T. brandy
3 C. apple slices

Combine flour, oatmeal, and sugar; cut in butter until mixture resembles coarse crumbs. Stir in nuts. Reserve 1 C. crumb mixture; press remaining mixture in bottom of greased 9x13-inch pan. Bake at 350° for 15 minutes. Combine cream cheese, marshmallow creme and brandy until well blended. Stir in apple slices. Spoon over crust; sprinkle with reserved crumb mixture. Bake at 350° for 25 minutes. Serve warm or cold.

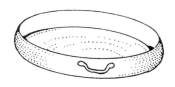

## APPLE CRISP

4 C. tart cooking apples  
¼ C. white sugar  
1 tsp. cinnamon  

¾ C. brown sugar  
½ C. flour  
¼ C. melted butter  

Slice apples into a greased 9-inch square pan; sprinkle with white sugar and cinnamon. Mix last 3 ingredients until crumbly. Spread over apples. Bake at 375° for 30-35 minutes.

## RAW APPLE BARS

MIX TOGETHER:  
2 C. sugar  
½ C. oil  

1 tsp. vanilla  
2 eggs  

Beat well.

ADD:  
4 C. sliced apples  
2 C. flour  
½ C. nuts  

1 tsp. salt  
1 tsp. soda  
1 tsp. cinnamon  

Beat well. Pour on greased and floured 11x16-inch sheet.

TOPPING:  
¼ C. brown sugar  
1 tsp. nuts  

¼ C. white sugar  

Sprinkle this mixture on top of dough. Bake at 350° for 45 minutes.

## PASTRY APPLE BARS

2½ C. sifted flour
1 tsp. salt
1 C. shortening
1 egg yolk

¾ C. sugar
1 tsp. cinnamon
1 egg white
1 C. corn flakes
8-10 tart apples (peeled and sliced; 6 C.)

Combine flour and salt. Cut in shortening. Beat egg yolk in ⅔ C. measuring cup, add milk to make ⅔ C. liquid. Mix well. Stir into flour mixture. On floured surface, roll ½ dough to fit bottom and up sides of 15½x10½x1-inch baking pan. Sprinkle with corn flakes and then add sliced apples. Combine sugar and cinnamon and sprinkle over apples. Roll out second ½ of dough to cover apples. Seal edges and slit top. Beat egg white until frothy and brush on crust. Bake in 375° oven for 50 minutes or until golden brown. Drizzle powdered sugar icing over warm bars. Cut; makes 3 dozen bars.

## SPICY APPLE COOKIES

½ C. butter
1 C. sugar
⅓ C. brown sugar
1 egg
2 C. flour

1 tsp. salt
1 tsp. soda
1 tsp. cinnamon
¼ C. milk
2 apples (finely chopped)
1 C. raisins

Cream butter; add sugars while continuing to cream. Add egg, beaten well. Mix flour, soda, cinnamon, and salt together. Add remaining ingredients; mix well. Drop by spoonful on lightly greased cookie sheet about 2-inches apart. Bake at 400° for 10-12 minutes or until golden brown.

## APPLESAUCE CAKE

¾ C. raisins
½ C. boiling water
½ C. shortening
1½ C. brown sugar
1 tsp. salt
½ tsp. cinnamon

½ tsp. cloves
½ tsp. allspice
2 eggs
1½ C. applesauce
2½ C. flour
¾ C. chopped walnuts
2 tsp. soda

Combine raisins and boiling water; set aside. In mixing bowl, beat shortening, sugar, salt, spices, and eggs until creamy. Scrape bowl often. Blend in applesauce. Beat; then add flour and nuts. Stir soda into undrained raisins; add to batter and mix by hand well. Pour into greased and floured 9x13-inch pan. Bake at 350° for 40-45 minutes. Frost with Never Fail Caramel Frosting.

NEVER FAIL CARAMEL FROSTING:
7½ T. brown sugar
7½ T. milk
4½ T. sugar

3 T. butter
9 large marshmallows
1½ C powdered sugar

Combine brown sugar, milk, sugar, and butter. Boil 2 minutes. Remove from heat; add marshmallows. Cool. When cool, add powdered sugar. Beat well. Spread on cake.

## HOT SPICED APPLE CIDER

2 qts. apple cider or juice
1¼ qts. cranberry juice
1 tsp. salt

4 cinnamon sticks
½ tsp. whole cloves
¼ C. brown sugar

Mix juices together. Pour into 30-cup coffee maker. Into filter mix cinnamon sticks, cloves, salt, and brown sugar. Perk as you would coffee. Delicious!

## APPLE CRUNCH

3-4 C. apples (sliced)
½-¾ C. sugar
1 T. flour
Sprinkle of cinnamon
⅓ C. butter (melted)
¾ C. oatmeal

¾ C. flour
¾ C. brown sugar
¼ tsp. salt
½ tsp. soda
½ tsp. baking powder

Combine apples, sugar, flour, and cinnamon and place in bottom of greased 8x8-inch pan. Mix together butter, oatmeal, flour, brown sugar, salt, soda, and baking powder. Sprinkle on top of apples. Bake at 350° for 50 minutes.

## Apple Rum

2⅔ C. apple cider
4 T. brown sugar
4 tsp. butter

4 cinnamon sticks
6 oz. rum
Dash of nutmeg

Into each of 4 mugs, combine ⅔ C. cider, 1 cinnamon stick, and 1 T. brown sugar. Heat at high in microwave for 3½ to 4 minutes. For each drink, stir in 1½ oz. rum and top with 1 tsp. butter and dash of nutmeg. Makes 4 servings. (For 2 servings, halve all ingredients and microwave for 1½ to 2 minutes.)

## APPLE SLICES

MIX AS FOR PIE:
2 C. flour
1 C. shortening
1 tsp. salt

ADD:
1 egg yolk
⅓ C. milk

Divide dough and roll out to fit cookie sheet. Put 4 C. sliced apples, cinnamon, sugar, 4 T. butter on crust. Scatter ¼ C. flour over it. Put the other rolled out dough on top. Bake at 400° for 15-20 minutes. Frost while hot with powdered sugar glaze.

## APPLE SALAD DRESSING

¼ C. vinegar
1 C. sugar
¾ C. water

2 eggs
1½ tsp. flour
1 tsp. vanilla

PLACE IN BOWL:
Fresh fruits (cut up)
Chopped nuts

Marshmallows

Combine all dressing ingredients in saucepan and cook until it becomes thick. Chill. (If too thick, thin with a little milk.) This is a good dressing for fresh fruits such as apples, oranges, and bananas, cut up small. Add marshmallows and chopped nuts.

## APPLE FRUIT SALAD

4 C. chopped apples  
8 oz. crushed pineapple  
2 C. salted Spanish peanuts  
8 oz. Cool Whip  

1 egg (beaten)  
2 T. vinegar  
½ C. sugar  
1 T. flour  

Drain pineapple. Combine pineapple juice, sugar, flour, and vinegar. Add beaten egg. Blend well. Cook on stove and stir until thick. Cool completely. Fold in Cool Whip. Add pineapple, apples, and peanuts. Chill.

## APPLE AND YAM CASSEROLE

Preheat oven to 325°. Combine 21 oz. can apple pie filling and 2 (17 oz. ea.) cans whole sweet potatoes, drained, in buttered 7x11-inch baking dish. Dot with 3 T. butter, sprinkle lightly with nutmeg and chopped nuts. Bake 30 minutes, until bubbly. Serves 8.

## Crab Apple Pickles

1 gal. apples (washed and pricked)  
5 C. sugar  
4 C. dark vinegar  
3 C. water  

2 sticks cinnamon  
1 T. whole allspice  
½ T. whole cloves  
Gauze bag  

In large kettle mix sugar, vinegar, and water. Tie spices in gauze bag and add to syrup mixture. Bring syrup to a boil and cook until sugar dissolves. Remove from heat and cool. When cooled add apples (whole) and return to heat. Simmer until apples are tender, but not mushy. Remove from heat and let stand 12-18 hours. Remove apples from syrup and pack into hot jars. Heat syrup and pour over apples. Seal with hot lids.

## APPLE BUTTER

¼ C. lemon juice  
¼ C. vanilla  
1 lb. brown sugar  
2½ lbs. white sugar  

Cinnamon to taste  
2 boxes pectin  
¼ C. butter  
2 pecks apples (your choice)  

Peel and slice apples. Place in large kettle. Add 2 sugars, vanilla, lemon juice, and butter. Heat. When hot, add pectin and continue cooking until thick. (Add cinnamon to taste.) Place in covered container and refrigerate.

# Mashed Potatoes with Apple

Mashed potatoes, leftover   Buttered crumbs
½ as much cooked apples as potatoes

Measure mashed potatoes. Mash cooked apples, drain well and measure ½ the amount of mashed potatoes. Stir to combine. Top with buttered crumbs. Place in heat-proof bowl and warm in 350° oven or heat in microwave.

# Iced Apple Brownies

3 eggs
1¾ C. sugar
1 C. oil
1 tsp. cinnamon
1 C. chopped nuts

2 C. flour
1 tsp. salt
1 tsp. baking soda
1 C. finely chopped apples

Cream eggs with sugar and oil. Add cinnamon and nuts. Stir in dry ingredients and apples. Pour into a 9 x 13-inch pan (greased) and bake in 350° preheated oven for 35-40 minutes.

ICING:
¼ C. butter
1 (3 oz.) pkg. cream cheese
  (softened)

1 C. nuts (finely chopped)
½ box (8 oz.) powdered sugar
1 tsp. vanilla

Beat all ingredients with mixer until smooth. Spread on cooled brownies.

## APPLESAUCE CAKE

½ C. shortening
1 C. sugar
1 egg
1½ C. flour
1 tsp. soda
1 tsp. salt

1½ tsp. cinnamon
½ tsp. cloves
1 C. hot applesauce
1 C. raisins
1 C. chopped nuts

Cream together shortening, sugar and egg. Add flour, soda, salt, cinnamon and cloves alternately with applesauce. Add raisins and nuts. Bake at 350° until done.

## Apple Pudding Bar

1 C. flour
1½ T. sugar
1 large pkg. instant vanilla pudding
1 large Cool Whip

1 stick oleo
4 C. apples (sliced)
Sugar and cinnamon

Mix flour, oleo, and sugar. Put in 9 x 13-inch greased pan. Press down. Add 4 C. apples. Sprinkle with sugar and cinnamon. Bake 45 minutes at 375°. Cool. Mix vanilla pudding as per instructions on box. Spread on cooled mixture. Top with Cool Whip. Refrigerate.

## Apple Crunch

1 C. white sugar
¼ tsp. salt
2 T. flour
1 tsp. cinnamon
8 heaping C. apples
  (sliced & peeled)

1 C. oatmeal
1 C. brown sugar
¼ tsp. soda
¼ tsp. baking powder
1 C. flour
½ C. margarine (melted)

Mix white sugar, salt, 2 T. flour, cinnamon, and apples together. Spread in 9 x 13-inch well-greased pan. Mix oatmeal, brown sugar, soda, baking powder, 1 C. flour, and melted margarine. Sprinkle on top of apple mixture. Bake at 350° until done, approximately 45 minutes. This is also good made with fresh rhubarb.

## Apple Crisp

3-4 medium sized apples
½ tsp. cinnamon
½ C. brown sugar
½ C. oatmeal

½ C. sugar
½ C. butter or margarine
½ C. flour

Slice apples into 8 x 8-inch buttered pan. Sprinkle with sugar and cinnamon. Place over low heat and partially cook while mixing topping. Mix remaining ingredients together and spread over apples. Bake at 350° for about 25 minutes or until browned. This is easy to remember, because everything is measured by halves.

# Apple Crisp

4 C. sliced apples  
1 C. flour  
¾ C. brown sugar (packed)  
1 T. cinnamon  
1 T. nutmeg  
½ C. butter or margarine

Put apples into a buttered 1½-qt. baking dish or pan. Blend flour, sugar, cinnamon, nutmeg, and margarine. Sprinkle over apples. May sprinkle sugar and cinnamon between layers of apples. Bake for 30 to 35 minutes until apples are tender and topping is crusty and brown. M-m-m good!

# Apple Crisp Pudding

6 or 8 apples (sliced)  
1 tsp. cinnamon  
½ C. wataer  
½ C. butter  
1 C. sugar  
¾ C. flour

Butter casserole and add apples, water, and cinnamon. Work together sugar, flour, and butter until crumbly; sprinkle over apple mixture. Bake, uncovered at 350° for 30 minutes.

## Apple Crisp

1 C. flour
1 C. sugar
¾ tsp. baking powder
¼ tsp. salt

1 egg
Sliced apples
3 T. margarine
Cinnamon

Mix flour, sugar, baking powder, and salt. Beat egg. Cut into mixture until crumbly. Pour over top of sliced apples in square pan. Melt margarine and pour over top. Sprinkle with cinnamon. Bake at 350° for 35 to 40 minutes.

## Apple Bread

1 stick oleo
1 tsp. vanilla
1 C. sugar
2 eggs
2 T. milk
1 C. apples (chopped fine)

½ C. nuts
2 C. flour
2½ tsp. baking powder
1 tsp. cinnamon
½ tsp. salt

GLAZE:
½ C. powdered sugar
1 T. warm water

2 T. salad oil
½ tsp. cinnamon

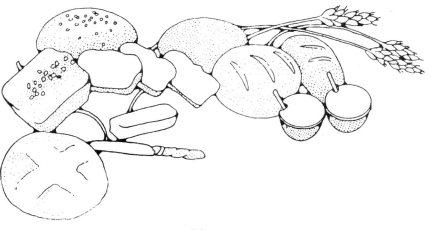

# Red Apple Cole Slaw

1½ C. unpeeled, diced red apple  
2 C. cabbage, finely shredded  
¼ C. seedless raisins  
½ C. chopped nuts  
½ C. coleslaw dressing

Combine all ingredients and toss well to distribute dressing. Chill if not using right away.

# Fried Apples

4 red-skinned apples  
¼ C. raisins  
2 T. oil or bacon fat

Core and slice apples. Fry apples in oil or bacon fat 10 to 15 minutes; add raisins and stir. Serve with pork chops or ham.

If you're calorie counting, "fry" apples with 2 to 3 T. water. Unbelievably good!

# Apple Cinnamon Pie

6 to 8 large apples, peeled & sliced*  
1 to 1½ C. sugar  
2 T. flour  
¾ tsp. cinnamon  
Unbaked pastry for 2-crust pie

Combine sugar, flour, and cinnamon and stir in with apples. Fill bottom crust with mixture. Top with crust and cut slits so steam can escape. Bake at 400° for 50 to 60 minutes. Check to be sure apples are done.

*Two (1 lb. 4 oz.) cans of sliced pie apples can be used instead of fresh apples.

# Ginger Pears

4 pears, peeled & cored  
1 T. lemon juice  
4 T. brown sugar  

4 T. white wine  
1 slice fresh ginger root or  
½ tsp. ground ginger  

Place pear halves in a baking dish and sprinkle with the lemon juice and sugar. Smash the ginger with the side of a cleaver and place in baking dish or sprinkle with ground ginger. Add the wine and bake at 350° for 25 to 30 minutes. Serve hot or cold. Serves 4.

## Pear Bread

½ C. butter  
1 C. sugar  
3 eggs  
2½ C. flour  
½ C. sour cream or yogurt  
1 C. pears (peeled & chopped)  

1 tsp. almond extract  
½ tsp. salt  
1 tsp. baking powder  
¼ tsp. soda  
½ tsp. nutmeg  

Cream butter and sugar; add eggs and 1 at a time. Add dry ingredients. Fold in sour cream or yogurt and almond extract until well blended. Fold in chopped pears. Pour into large loaf pan that has been well oiled. Bake at 350° for 1 hour.

## Dip for Fruit Slices

1 (8 oz.) pkg. cream cheese, softened
½ C. brown sugar
¼ C. powdered sugar
¾ tsp. vanilla

Blend all ingredients together using the electric mixer. Store covered overnight in refrigerator. Delicious with apple slices.

Adding a little vinegar to water used to boil hard-cooked eggs will keep them from "running" if a shell cracks.

## BAKED PEARS

Pears  
Butter  
Brown sugar  

⅓ C. tapioca  
1 C. water  
Maraschino cherries

Wash as many pears as you need to prepare. Take out cores and stems and cut pears in half. Put cut side up in baking dish. Slice a thin slice of butter and cut it into 4 pieces. Put one small square in each pear center. Sprinkle 1 tsp. brown sugar over the half (more if you want it sweeter). Soak ⅓ C. of tapioca in the cup of water and pour over and around the pears. Put cherry in the center of each pear half with some cherry juice. Bake slowly (325°) until pears are cooked. Pears can be turned once in juice that will be produced. If too dry, cover with foil. Depending on the type of pear, bake about 1-1½ hours. A scoop of vanilla ice cream goes nicely with the pears.

# Fruit Dip

1 (8 oz.) pkg. softened cream cheese  
1 T. lemon juice  
1 (7 oz.) jar marshmallow creme

Mix with electric mixer. Chill and serve. Good with any fresh fruit.

## PEACH COBBLER

4 C. sliced peaches

2 T. oleo or butter  
½ C. sugar  
½ C. water  
2 T. flour  
1 tsp. cinnamon (I use ½ tsp.)

2 C. flour  
3 tsp. baking powder  
¼ tsp. salt  
4 T. shortening  
¾ C. milk  
1 egg

Blend peaches, oleo, sugar, water, flour, and cinnamon. Pour into shallow greased pan. To make crust: Mix flour, baking powder, salt. Work in shortening. Add milk and egg. Spread soft mixture over peaches. Leave holes in top. Bake ½ hour or more in moderate oven. Pour over this syrup when taken from oven: ½ C. sugar and ½ C. water. Bring to boil 2 minutes. Pour over cobbler and bake 5 minutes more. Serve warm. Good with milk or scoop ice cream on top.

## FRESH PEACH PIE

1 (3 oz.) pkg. peach or apricot Jello
2 T. (rounded) cornstarch
⅓ C. sugar
1 C. water

3 C. diced peaches
1 C. flour
1 stick margarine
½ C. chopped nuts

Heat first 4 ingredients until thick or clear. Cool. Add 3 C. diced peaches. For the crust mix flour, margarine, and nuts and press in a 9-inch pie pan. Bake crust for 20 minutes at 350°. Cool crust and pour Jello mixture in it. Top with Cool Whip and refrigerate.

## Fresh Peach Pie

5 T. peach Jello
3 T. cornstarch
1 C. sugar

1 C. water (if peaches are real juicy use less water)
3 fresh peaches
1 baked 9-inch pie shell

Boil Jello, cornstarch, sugar and water until thick and clear. Cool. Line pie shell with a little of the pudding. Slice fresh peaches into pie shell. Cover with pudding. Refrigerate. (For strawberry pie use strawberry Jello.)

## Fried Peaches

8 large peaches  
3 T. butter  

1 T. dark brown sugar  
Vanilla ice cream  

Peel, halve the peaches, and remove seeds. Cut thin slice off rounded side of peach to level. Melt the butter in a large skillet and put in the peaches, leveled side down. Fill hollows with brown sugar. Simmer, covered until just beginning to get soft. Serve with a scoop of vanilla ice on top of each peach half.

## Peach Cheese Dessert

1 C. + 2 T. flour  
1 C. chopped nuts  
¼ C. brown sugar  
1 stick margarine (melted)  
8 oz. cream cheese  
½ C. powdered sugar  
1 tsp. vanilla  

2 C. Cool Whip  
1½ C. water  
¼ C. sugar  
2 T. cornstarch  
3 oz. peach Jello  
4 C. peaches (sliced)  

Mix the first 4 ingredients and press in 9×13-inch pan. Bake 15 minutes at 350°. Cool. Mix softened cheese, sugar, vanilla, and Cool Whip. Spread over crust and chill. Cook water, sugar, and cornstarch until thick. Add Jello and sliced peaches. Pour over and chill.

## FRESH PEACH PIE

PIE CRUST:
1½ C. flour
½ C. corn oil
½ tsp. salt
2 T. sugar
2 T. milk

PEACH FILLING:
½ C. sugar
3 T. cornstarch
2 T. white Karo syrup
1 C. water
3 T. peach gelatin (dry)
7 ripe peaches

Combine pie crust ingredients; mix with electric mixer. Press into pie pan and bake at 350° for 15-20 minutes. Cool.

For Filling: Combine sugar, syrup, cornstarch, and water. Bring to boil and stir until thick and clear. Remove from heat and add dry gelatin. Mix well; cool. Skin peaches and slice thin. Add fresh peaches and put in crust. Chill, serve with whipped cream. This pie doesn't keep well. Should be made the day you are going to serve it.

## CREAMY PEACH PIE

3 C. peeled, sliced peaches
¾ C. white sugar
¼ C. flour
¼ tsp. salt
¼ tsp. nutmeg
1 C. whipping cream

Prepare 9-inch unbaked pastry shell. Add above mixture to peaches. Toss lightly. Turn into pastry shell. Pour 1 C. whipping cream over top of peaches. Bake at 400° for 35-45 minutes.

## FRUIT DIP

7 oz. marshmallow creme
8 oz. cream cheese
Dash of food coloring
Dash of ginger

Mix all together and serve with fresh fruit.

## GOLDEN PEACH PIE

2 (1 lb. ea.) cans peaches (drained)
⅓ C. peach juice
½ C. sugar
¼ tsp. nutmeg
2 T. flour
2 T. lemon juice
1/8 tsp. almond extract
2 T. butter

Cook and stir until mixture thickens. Pour into 9-inch pie crust. Add top crust. Bake at 375° for 40-45 minutes

# Peach Upside Down Cake

1 C. brown sugar
½ C. Crisco
1 C. white sugar
½ C. Crisco
1 egg
½ C. milk

1½ C. flour
2 tsp. baking powder
¼ tsp. salt
1 tsp. vanilla
Peaches

Place brown sugar, and Crisco in large iron skillet. When melted, place in pan as many sliced or halved peaches as possible. Pour the cake mixture over peaches. Bake at 350° for 40 minutes. Turn out on cake plate, peach side up. May be served with whipped cream.

# Peach Apple Conserve

2 C. peeled & chopped peaches
2 C. unpeeled & chopped red apple

2 T. maraschino cherries, minced
½ C. fresh lemon juice
3 C. sugar

Combine all ingredients in heavy saucepan. Cook over low heat 20 to 30 minutes until apples are transparent. Pour into hot sterilized jars and seal at once.

# Gotcha Peach Punch

8 ripe peaches
2 bottles dry white wine (chilled)
4 maraschino cherries

Pour boiling water over the peaches and let stand about ¾ minute. Remove peaches with slotted spoon. Skins will peel off readily.

Put peaches into a clear glass pitcher. Pour wine over peaches immediately to prevent discoloration. Refrigerate for 3 hours so peaches will flavor the wine.

To serve, place pitcher on the dinner table. As wine is poured, add additional chilled wine.

Peaches can be served sliced over vanilla pudding or ice cream for an elegant dessert.

## FRUIT PIZZA

1 pkg. sugar cookie dough
1 pkg. cream cheese
⅓ C. sugar
1 tsp. vanilla

Fresh fruit (strawberries, grapes, peaches, bananas, pineapple)
½ C. apricot preserves
2 T. water

Slice and arrange cookie dough on a pizza pan. Bake as directed; cool. Cream together the cream cheese, sugar and vanilla. Frost cookies. Top with fruit. Combine the preserves and water, drizzle over the top making sure the bananas are covered so they don't turn dark.

## Fresh Peach Pie

1 C. sugar
1 C. water

2 T. cornstarch

Boil above, stirring constantly until thick and clear. Then dissolve ½ pkg. (3 T.) peach Jello in above mixture. Let this cool, but not set. Put a coat of this mixture in baked pie shell, slice peaches in a layer, add more Jello mixture, then layer of peaches and end with Jello mixture. Be sure peaches are covered. Set in refrigerator. Put whipped cream on top. Use Fruit Fresh in peaches to keep color if desired.

## PEACHIE PUDDING BAKE

8 medium peaches
¾ C. brown sugar

1 C. Bisquick
¼ tsp. nutmeg

Peel peaches and slice into a large bowl. Sprinkle evenly with sugar and let stand 10 minutes. Stir in Bisquick and nutmeg. Turn fruit into greased non-metallic baking dish. Micro-cook on roast temperature for 15 minutes. Serve with ice cream. Makes 6 servings.

## Grandma's Cream Peach Pie

1 unbaked pie shell

2 skimpy layers sliced peaches

Mix ¾ C. sugar and 3 T. flour. Sprinkle this mixture over peaches. Pour ¾ C. Half and Half over all. Sprinkle nutmeg on top. Bake at 350° for 45 minutes.

## You Name It

1 (3½ oz.) pkg. instant vanilla pudding
1½ C. milk
1 (7 oz.) pkg. small jelly rolls

2½ C. whipping topping, divided
3 C. sliced strawberries (reserve 3 whole ones with leaves for garnish)

Prepare the instant pudding with 1½ C. milk. Allow to sit for 10 minutes. Fold in 1 cup of the whipped topping.

Slice each jelly roll into 3 pieces; stand these around the sides of a glass bowl so the red rings show. Put the extra slices on bottom of the bowl.

Spread ½ of the strawberries over the slices in the bottom of the dish. Pour in the pudding mixture. Spread the remaining strawberries over the pudding; spread the rest of the topping over the strawberries. Garnish with reserved berries. Store in the refrigerator. Serves 8.

## FRESH STRAWBERRY PIE

1 C. sugar
1 C. water
2 T. cornstarch

2 T. white corn syrup
3 T. dry strawberry gelatin
1 qt. fresh strawberries

Combine sugar, water, cornstarch, and corn syrup. Cook until clear, add gelatin and cool to room temperature. Pour over strawberries in baked shell. Chill. Serve with whipped cream.

## Fruit Soup

3 pts. water
½ C. each of the following:
   pitted prunes, seedless raisins,
   dried apples, dried apricots,
   dried apples (all fruits chopped
   into ½-inch chunks, except raisins)
¼ C. raw rice
1 can pitted dark sweet cherries
   (undrained)
1 piece stick cinnamon
   (2 to 3 inches long)

Although called soup, it is usually served (hot or cold) as dessert. In large pan boil water, then gradually add rice. Boil, uncovered about 15 minutes until rice is soft. Reserve liquid after draining rice and return liquid to pan; bring to boil. Add dried fruits and cinnamon stick. Simmer for 30 minutes; remove from heat and add cherries with juice. Stir well; stir in rice and serve hot. It also may be served at room temperature or chilled, topped with whipped cream.

## Berry Delight

1½ C. graham cracker crumbs
⅓ C. butter (melted)
¼ C. sugar
3½ C. Cool Whip
2 pkgs. (4 serving size) vanilla
   instant pudding
3½ C. cold milk
¼ C. sugar
1 pkg. (8-oz.) cream cheese
   (softened)
2 T. milk
2 pt. strawberries (hulled
   and halved)

Combine crumbs, ¼ C. sugar and butter. Press into bottom of a 9x13-inch pan; chill. Beat cream cheese, ¼ C. sugar, and 2 T. milk until smooth. Fold in ½ of the Cool Whip; spread over crust. Arrange berries in an even layer. Using the cold milk prepare pudding as on package. Pour over the berries. Chill several hours. Shortly before serving spread rest of Cool Whip over pudding, garnish with remaining strawberries. You can also use bananas, peaches or pineapple, instead of the strawberries.

## Fruit Pizza

**CRUST:**
1 C. soft margarine
2 eggs
1 tsp. soda
2 tsp. cream of tartar
2 C. sugar
1 T. vanilla
2¾ C. flour

**FILLING:**
8 oz. cream cheese (softened)
1 tsp. vanilla
½ C. sugar

**FRUIT:**
2 bananas (soaked in pineapple juice)
8 oz. mandarin oranges
Kiwi fruit (opt.)
2 C. strawberries
1 C. chunk pineapple

**GLAZE:**
½ C. water
2 T. lemon juice
½ C. sugar
½ C. orange juice
1½ T. cornstarch
Dash salt

Mix up crust ingredients and put in large cookie sheet with sides. Bake at 350° for 12-17 minutes, until brown. Cool. Will deflate while cooling. Next mix up the filling ingredients and spread n cooled crust. Next arrange fruit on crust in circles. Cook glaze over medium heat until it boils. Cool 1 minute, it should be thick. Pour on fruit and refrigerate.

## STRAWBERRY ICE CREAM DELIGHT

1 (6 oz.) pkg. red Jello
1½ C. boiling water
1 C. canned pineapple juice

Ice cubes
1 C. vanilla ice cream
1 C. sliced, fresh strawberries

Completely dissolve Jello in water. Combine pineapple juice and ice cubes to make 2½ C. Add to Jello and stir until slightly thickened; remove unmelted ice. Measure 1 C. Jello and stir in ice cream, let stand to thicken if necessary. Add berries to remaining Jello and spoon about ⅓ into large bowl. Add ½ of creamy mixture. Repeat ending with fruit mixture. Chill until set.

## Fruited Dessert
*(Makes about 5 C. or 10 servings)*

2 (4-serving size ea.) pkgs. or
1 (8-serving size) pkg. Jello strawberry or strawberry-banana flavor*
1½ C. boiling water

1 C. cold canned pineapple juice*
Ice cubes
1 C. (½ pint) vanilla ice cream*
1 C. sliced or diced strawberries*

Completely dissolve gelatin in boiling water. Combine juice and ice cubes to make 2½ C. Add to gelatin, stirring until slightly thickened. Remove any unmelted ice. Measure 1 C. of the thickened gelatin and stir in ice cream. Let stand a few minutes to thicken even more. Stir fruit into remaining gelatin. Spoon about ⅓ of the fruited gelatin into individual dessert dishes or serving bowl. Spoon half of the creamy mixture over fruited gelatin. Repeat layers, then spoon remaining fruited gelatin on top. Chill until set. Garnish as desired.

## STRAWBERRY DELIGHT

1 prepared angel food cake

1 (3½ oz.) pkg. instant vanilla pudding
1 C. sour cream

Sliced strawberries

1 C. milk
8 oz. Cool Whip

Combine pudding and milk; add sour cream. Fold in Cool Whip. Break angel food cake into pieces and place in a 9x13-inch pan. Add layer of pudding mixture, then a layer of strawberries. Repeat until all pudding mixture is used.


## STRAWBERRY CREAM DIP

2 C. whole, ripe strawberries
2 T. honey

1½ C. low-fat creamed cottage cheese

Whirl all ingredients in a blender until completely smooth. Divide equally in four small glass dishes. Place each dish on lettuce-lined dinner plate and surround with your favorite combination of fresh fruit. Makes 4 servings.

## STRAWBERRY-RHUBARB STREUSEL PIE

2 C. chopped rhubarb
1 C. frozen strawberries
1½ C. sugar
½ C. flour
½ tsp. cinnamon
1 tsp. lemon juice
1 (9-inch) pastry shell (unbaked)

Combine all ingredients and mix well; pour into pie shell. Cover with topping (below).

TOPING:
¾ C. sugar
½ C. butter or margarine
2 C. flour

Mix until crumbly and sprinkle over pie. Bake 50 minutes in a 350°-375° oven.

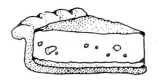

## STRAWBERRY-RHUBARB PIE

1½ C. sugar
¼ tsp. salt
3 C. (½-inch pieces) rhubarb
Pastry for 2-crust pie
¼ C. flour
¼ tsp. nutmeg
1 C. sliced strawberries
1 T. butter

Combine dry ingredients. Add fruit, mixing well. Let stand 20 minutes. Spoon in 9-inch pie crust. Dot with butter. Moisten edge, adjust top crust. Flute edge. Bake at 400° for 40-45 minutes or until done. Serve slightly warm.

## Strawberry Graham Dessert

15 graham crackers
¼ C. sugar
½ C. milk
¼ C. melted butter
½ C. chopped nuts (opt.)

½ lb. marshmallows
1 C. whipping cream
1 pt. sweetened strawberries
 (2 C.)

Roll graham crackers fine and add melted butter and sugar. Mix well and pat into greased baking dish. Reserve 2 T. cracker crumbs for top. Melt marshmallows in top of double boiler with milk. Cool this mixture. Add cream which has been whipped. Place layer of marshmallow mixture in dish and then the strawberries and nuts mixed lightly together. Add remaining marshmallow mixture and top with reserved cracker crumbs.

## STRAWBERRY SHORTCAKE

4 eggs
2 C. sugar
½ tsp. salt
2 tsp. vanilla

2 T. butter
1 C. boiling milk
2 C. flour
2 tsp. baking powder

Beat eggs until very light. Beat in sugar, salt and vanilla. Melt butter in milk. Sift and beat in flour and baking powder. Bake in greased pan 25-30 minutes at 350°. Serve with strawberries and cream.

## RED RASPBERRY PIE

Baked 8-inch or 9-inch crust

2 C. raspberries (add ½ C. sugar and let set until juice forms).

Use juice plus ½ C. sugar, 2 T. cornstarch and 2 C. water. Bring to a boil and add 1 (3 oz.) pkg. red raspberry gelatin. Put berries in the pie shell. Add cooled glaze and refrigerate a few hours before serving.

## BLACK RASPBERRY JELLY

3½ qts. berries            1 box pectin
6½ C. sugar

Clean berries and cook in a small amount of water (1-2 C.) until soft and mushy. Strain through a cloth to remove seeds. Measure juice and place 4½ C. juice in a large pan. Stir in 1 box pectin. Bring to a boil and add sugar. Bring to a hard boil, stirring constantly and boil hard 1-2 minutes. Remove from heat. Skim off foam and place in containers. Refrigerate. *May cover with paraffin wax.

## Fresh Blueberry Glaze Pie

1 C. sugar
3 T. cornstarch
¾ C. water
1 qt. fresh blueberries
1 T. lemon juice
1/8 tsp. salt
1 T. butter

1 (8 oz.) cream cheese (softened)
1 C. powdered sugar
1 tsp. lemon peel
  (finely shredded)
1 tsp. lemon juice
1 C. whipped cream
1 (9-inch) baked pastry shell
Whipped cream

In saucepan, combine sugar and cornstarch; stir in water. Add 1 C. blueberries. Cook and stir until thickened and bubbly. Stir in 1 T. lemon juice, butter and salt. Cool to room temperature. Fold in remaining blueberries. Cover and chill. Combine cream cheese, powdered sugar; stir in lemon peel and 1 tsp. lemon juice. Fold in 1 C. whipped cream. Spread on bottom and sides of baked pastry shell. Just before serving, pour blueberry mixture into pastry shell. Top with additional whipped cream.

## Glazed Blueberry Pie

1 (3 oz.) pkg. cream cheese
1 (9-inch) baked pie shell
4 C. fresh blueberries
½ C. water

¾ C. sugar
2 T. cornstarch
2 T. lemon juice

Soften cream cheese; spread in bottom of cooked pie shell. Fill with 3 C. blueberries. Combine 1 C. blueberries with water. Bring to boil, reduce heat and simmer 2 minutes. Strain and reserve juice. Combine sugar and cornstarch, then gradually add reserved juice. Cook, stirring constantly until thick and clear. Cool slightly, then add lemon juice. Pour over berries in pie shell. Chill. Serve with whipped cream.

## CONCORD GRAPE PIE

FILLING:
1½ qts. washed grapes
1 C. sugar
2 T. cornstarch

CRUST:
3 C. sifted flour
1 tsp. salt
1¼ C. solid shortening
1 egg (well beaten)
1 T. vinegar
4 T. cold water

Squeeze pulp out of grapes and set aside the skins. Sieve seeds out of pulp and combine skins and pulp in saucepan. Add sugar and cornstarch and mix well. Cook over low heat until thick; set aside to cool. Sift flour and salt into a bowl; add shortening and mix with pastry blender. In another bowl, combine the egg, vinegar, and water. Drizzle over flour mixture and mix until it holds together in a ball. Divide dough in half. Roll into 2 rounds 1/8-inch thick. Place one round into a 9-inch pie pan. Pour in grape mixture. Top with remaining crust, seal, and cut several slits in top. Bake at 400° for 30 minutes or until crust is done and pie is bubbly. Check your oven - this temperature may be too hot for your crust. (NOTE: I boil the grape pulp and this makes the seeds separate more easily.)

## Cherry Pudding

2 C. cherries  
1 C. sugar  
1 C. flour  
1 C. sugar  

1 tsp. baking powder  
½ C. milk  
2 T. butter  

Combine 2 C. cherries with 1 C. sugar. Let stand while preparing the following:

Flour  
1 C. sugar  

Baking powder

Add milk and 2 T. melted butter. Pour batter into 9-inch pan. Sprinkle cherries on top and bake at 350° for 45 minutes or until ndone in center.

## RASPBERRY COBBLER

1 stick margarine  
1 C. sugar  
1 C. flour  

¾ C. milk  
1½ tsp. baking powder  

Melt margarine in 9x13-inch cake pan. Mix in bowl 1 C. sugar, 1 C. flour, ¾ C. milk, 1½ tsp. baking powder. Pour over margarine. DO NOT MIX. Put raspberries over this. Use fresh or frozen berries. Bake 375° about 30 minutes or until brown on top. (Any cooked fruit can be used in place of fresh berries.) Add cinnamon to apples; nutmeg to peaches.

# Glaze Strawberry Pie

1 (9-inch) baked crust

Fill baked crust with choice evenly sized strawberries placed with points up.

1 C. sugar
1 C. water
2 T. cornstarch
½ C. water

Boil sugar and water for 5 minutes. Mix cornstarch and water. Add to boiling syrup, stirring for 3 minutes. Add 3 T. strawberry Jello and stir well. Pour over fresh berries in crust. When cool, maybe decorated with whipped cream or Cool Whip; refrigerate.

# Applesauce-Cinnamon Salad

2 C. applesauce
½ C. cinnamon red hots
3 C. hot water
2 pkgs. (3 oz. ea.) cherry Jello
1 (8 oz.) pkg. cream cheese (softened)
4 T. milk
2 T. mayonnaise

Do ½ Jello mixture at a time. Boil red hots in water until dissolved. Pour into Jello and let cool; add applesauce. Put in a 9x12-inch pan and reefrigerate until set. Mix cream cheese, milk, and mayonnaise. Put on top of set Jello and let set. Make second half of Jello mixture and put on top of cream cheese.

## BERRIED DELIGHT

1½ C. graham cracker crumbs
¼ C. sugar
⅓ C. butter or margarine (softened)
1 (8 oz.) pkg. cream cheese (softened)
¼ C. sugar
2 T. milk

3½ C. (8 oz.) Cool Whip whipped topping (thawed)
2 (4-serving size ea.) pkgs. Jello vanilla flavor instant pudding and pie filling*
3½ C. cold milk
2 pints strawberries (hulled and halved)

*Or use other Jello instant pudding flavors with other fruits, such as bananas with banana cream or butterscotch, sliced pineapple with pineapple cream, or others. Combine crumbs and sugar. Mix in butter. Press firmly on bottom of a 9x13-inch pan. Bake, if desired, at 375° for about 8 minutes; cool. Beat cream cheese with sugar and 2 T. milk until smooth. Fold in half of the whipped topping. Spread over crust. Arrange strawberries in even layer. Prepare pudding with the cold milk as directed on package. Pour over berries. Chill several hours or overnight. Shortly before serving, spread remaining whipped topping over pudding. Garnish with additional strawberries, if desired; chill. Makes 15 servings.

## BLUEBERRY BUCKLE

2 C. sifted flour
2 tsp. baking powder
½ tsp. salt
¾ C. sugar

¼ C. butter or margarine
1 egg
2 C. fresh blueberries
½ C. milk

Sift together flour, baking powder, and salt. Cream butter and sugar, beat in egg. Add flour mixture and milk. Stir until dry ingredients are dampened. Fold in blueberries. Turn into 9-inch square cake pan. Sprinkle with topping (below). Bake in preheated oven, 325°, until tester comes out dry, 45-50 minutes. Loosen edges; let cool partly. Cut in squares in pan and serve warm.

TOPPING: Mix together:
½ C. sugar
½ tsp. cinnamon

⅓ C. unsifted flour
¼ C. soft butter or margarine

Mix until mixture looks like coarse crumbs.

## Blueberry Muffins

*1 C. sugar*
*½ C. oleo*
*2 eggs*
*½ C. milk*
*2 C. flour*

*½ tsp. cinnamon*
*2 tsp. baking powder*
*½ tsp. salt*
*1-1½ C. frozen or fresh blueberries*

Cream sugar, eggs, oleo. Combine remaining ingredients but fold blueberries in last. Bake at 375° for 20-25 minutes.

## MICHIGAN BLUEBERRY BUCKLE

2 C. flour  
¼ C. sugar  
2½ tsp. baking powder  
¾ tsp. salt  
¼ C. shortening  

¾ C. milk  
1 egg  
2 C. well drained blueberries  
(or frozen)  

Blend all ingredients except blueberries and topping, beat vigorously ½ minute. Stir in blueberries. Spread in 9x13-inch greased pan or 7x11-inch pan.

TOPPING:  
Mix ½ C. sugar  
⅓ C. flour  
1 tsp. cinnamon  
¼ C. soft butter  

Sprinkle over batter in pan. Bake at 375° for 45-50 minutes.

# Cream Fruit Pops

3 C. fruit juice, any kind  
Sugar, if juice is tart  
12 T. cream  
Sticks (tongue depressors)  

Combine ingredients. If sugar is used, stir to dissolve. Pour into 2 plastic popsicle molds, or pour into narrow jelly glasses. When "pops" are slushy, insert sticks. This can also be frozen in ice cube trays with toothpicks for "holders".

# Chewy Fruit Leather

5 to 7 very ripe pears, strawberries, plums or apricots

⅓ C. sugar

Peel and chop fruit to measure 5 cups. Put fruit and sugar in a heavy saucepan and bring mixture to a boil; stir to completely dissolve sugar. Puree in blender and cool. Cover cookie pan with plastic wrap and spread puree to 1/8-inch thickness.

Set pan on a level surface in the sunlight. Use a screen cover to avoid insects. Dry 20 to 24 hours. Bring sheet in at sunset and put out the next day. Fruit leather can be dried in a 150° oven with door open. When plastic peels off evenly, fruit is dry.

To store, place leather on a plastic sheet and roll up tightly. It will keep 3 to 4 weeks at room temperature; about 4 weeks in the refrigerator; about 1 year in the freezer.

To eat, just pull off chunks or slice into strips. Kids love this for an after-school treat.

# Orange Juice Squeezin's

Freshly squeezed or frozen orange juice (prepared as directed)

Chilled champagne

Pour equal parts of orange juice and champagne into punch bowl or glasses and serve. Great for brunch!

## Cherry Pudding Cobbler

1 C. sugar
¼ C. shortening
2 C. flour
2 tsp. baking powder
1 tsp. almond

1 C. milk
1 C. sugar
2 C. warm water
Lump of butter
2 C. cherries

Cream sugar and shortening; add flour sifted with baking powder. Add 1 C. milk to make a little stiffer batter than cake mixture. Spread in a greased shallow 9x13-inch pan. Pour on the 2 C. of cherries (or other canned or fresh fruit), 1 C. sugar, 2 C. warm water, and a lump of butter. Bake in moderate oven (350°-375°) for 45 minutes or until brown.

## CRANBERRY GELATIN SALAD

1 (1 lb.) pkg. cranberries (ground fine)
2 C. sugar
1 (20 oz.) can crushed pineapple

Combine and let stand for 1 hour. Dissolve 2 boxes raspberry Jello in 2 C. boiling water. Add to cranberry mixture and cool until congealed. Fork up and add ½ lb. miniature marshmallows, 1 pkg. Dream Whip, or 9 oz. Cool Whip and 1 C. nuts (chopped). Spread into a 9x13-inch cake pan.

## Cranberry Salad

1 C. sugar
¾ lbs. marshmallows
2 C. cranberries (1 lb.) (ground)
1 pt. whipping cream
1 C. drained pineapple (crushed)

Wash cranberries and grind. Combine cranberries and sugar; let stand for 2 hours (keep cool). Whip cream and combine with marshmallows. Also let stand for 2 hours (be sure to keep cool). After 2 hours put all together and add drained pineapple. Put in pan or mold. Let stand all night in refrigerator.

## Apple-Cherry Drink

2 pkgs. cherry Kool-Aid  
½ C. sugar  
1 (6 oz.) can frozen lemonade  

1½ qt. apple juice  
2¼ qt. ice water  

Mix and serve cold. Makes 1 gallon.

## Fruit Cobbler

3 C. diced rhubarb  
   (or any fruit)  
1¾ C. sugar  
3 T. butter  
¼ C. shortening  

1½ C. flour  
¼ tsp. salt  
3 tsp. baking powder  
1 beaten egg  
½ C. milk  

Put fruit in 8x12-inch greased pan and sprinkle with 1 C. sugar and dot with butter. Heat in 350° oven. Sift dry ingredients with the ¾ C. sugar. Cut in shortening and add egg which has been mixed with milk. Pour over hot fruit and bake at 350° for 50 minutes. May serve with a hot vanilla or cinnamon sauce.

# Fruit Cobbler

½ C. margarine
1 C. biscuit mix
¾ C. sugar
1 C. milk

4 C. fruit, drained
(cherries, blackberries, peaches
or any combination)

Melt margarine in 10x6-inch baking dish. Add biscuit mix, sugar, and milk. Stir to blend. Top with fruit. If unsweetened cherries are used, increase sugar ⅓ to ½ cup. Bake at 375° for 30 to 40 minutes. Top should be a light golden brown. Good topped with whipped cream or ice cream. Serves 5 to 6.

## Cherry Coffee Cake

1 C. oleo (creamed)
1 C. white sugar
¾ C. brown sugar
4 eggs
3 C. flour

1½ tsp. baking powder
¼ tsp. salt
1½ tsp. vanilla
1 can cherry pie filling

Mix creamed oleo and sugars well. Add eggs, 1 at a time. Combine dry ingredients and add to oleo-sugar-egg mixture. Save 2 C. batter. Spread remaining batter on greased cookie sheet. Spread cherry pie filling on top of batter to ½-inch of sides. Drop remaining batter by teaspoon on top of cherry filling at random. Bake at 350° for approximately 45 minutes until golden brown. Watch so doesn't get too brown. Drizzle white frosting over top of coffee cake. You can use other pie filling, if desired.

## Cranberry Salad

1 lb. fresh cranberries
1½ C. sugar
1 pt. real whipping cream
1 C. chopped pecans
1 C. sliced Tokay grapes

Grind cranberries and stir in sugar. Refrigerate overnight and stir a couple of times. Put cranberries in colander and drain the liquid off as much as possible. Whip cream until very stiff. Add cranberries, pecans, and grapes; refrigerate.

## Concord Grape Pie

4 C. Concord red grapes
1 C. sugar
½ C. flour

¼ tsp. salt
1 T. lemon juice
2 T. oleo (melted)

TOPPING:
½ C. flour
¼ C. oleo

½ C. sugar

9-inch unbaked pastry shell

Slip skins from grapes and set skins aside. Bring pulp to a boil then reduce heat and simmer uncovered for 5 minutes. Sieve pulp to remove seeds and add grape skins. In mixing bowl, mix sugar, flour and salt. Then add lemon juice and oleo and grape mixture. Pour grape mixture into pastry shell. Bake at 400° for 25 minutes. Then mix topping and sprinkle atop the pie and bake 15 minutes more.

## FRESH GOOSEBERRY PIE

1½ C. gooseberries  
1½ C. sugar  

3 T. flour  
1½ C. cream or Half and Half  

Wash and cut gooseberries in half. Blend sugar and flour. Add cream or Half and Half. Stir in gooseberries. Pour into unbaked pie crust and bake.

## Blackcap Berry Pie

2 C. fresh berries  
  (washed & drained)  
1½ C. sugar  
3 T. flour  

A double crust for reg. pie  
(or a crumb crust can be  
used for top)  

Pour sugar over berries. Add flour and mix well; let stand while mixing the double crust as for regular pies. Pour pie mixture into crust and bake at 375° for 10 minutes. Then bake pie at 350° for 45 minutes. (May be served with whipped cream or ice cream.)

## RAINBOW ROLL

1 egg  
1 C. icing sugar  
2 oz. unsweetened chocolate  
  (melted)  

½ pkg. colored marshmallows  
  (mini)  
Desired nuts, cherries  
Coconut  

Mix all ingredients. Spread coconut on waxed paper. Roll chocolate mix in it. Chill in refrigerator. Slice when fully chilled or ready to serve. Makes 2 rolls.

# VEGETABLES

| | |
|---|---|
| Anise Carrots .......................... 70 | Creamed Vegetables ................... 142 |
| Bean Salad ............................ 124 | Crock Pot Chicken & Veggies ........... 68 |
| Beat Up Taters ........................ 86 | Cucumber & Tomato Vinaigrette ......... 93 |
| Beet Pickles .......................... 130 | Cucumber Salad ....................... 93 |
| Belle Pepper Sauce .................... 121 | Cucumber Salad ....................... 94 |
| Broccoli Casserole .................... 108 | Cucumber Salad ....................... 94 |
| Broccoli/Cauliflower Salad .............. 126 | Cucumber Salad ....................... 95 |
| Broccoli/Cauliflower Salad .............. 143 | Cucumber Salad ....................... 95 |
| Broccoli/Cauliflower Salad .............. 146 | Cucumber Snacks ...................... 95 |
| Broccoli/Cauliflower Salad .............. 148 | Cucumbers with Sour Cream ............ 96 |
| Broccoli/Cauliflower Soup .............. 145 | Curry Dip for Vegetables ............... 101 |
| Broccoli Delight Salad .................. 108 | Dad's Rhubarb Cake .................. 115 |
| Broccoli Sesame ...................... 107 | Dill Pickles .......................... 149 |
| Browned Paprika Potatoes .............. 90 | Dill Pickles .......................... 149 |
| Bud's Spuds .......................... 89 | Dilly Zucchini ........................ 107 |
| Butter Beans ......................... 125 | Dot's Rhubarb Torte .................. 120 |
| Cabbage Salad ........................ 98 | Easy Dill Pickles ...................... 96 |
| Caramel Corn ......................... 77 | Eggplant Creole ...................... 131 |
| Carrot Cake .......................... 69 | Freezer Corn ......................... 78 |
| Carrot Cake .......................... 70 | French Fried Onion Rings .............. 133 |
| Carrot Casserole ...................... 111 | Fresh Broccoli Salad ................... 109 |
| Carrot Coleslaw ....................... 67 | Frosted Pumpkin Bars .................. 151 |
| Carrot Loaf ........................... 69 | Frozen Pumpkin Pie ................... 152 |
| Carrot Salad .......................... 67 | Garden Macaroni Salad ................ 92 |
| Carrot, Zucchini Bars .................. 71 | Garden Patch Dip ..................... 98 |
| Cauliflower Salad ..................... 127 | Garden Salad ........................ 148 |
| Cauliflower-Broccoli Salad .............. 101 | Garden Salad ........................ 91 |
| Cauliflower/Broccoli Salad .............. 146 | Garden Tomato Salad ................. 75 |
| Cauliflower/Cheese Sauce .............. 128 | Garlic Potatoes ....................... 90 |
| Cauliflower & Pea Salad ................ 127 | Green Bean Bake ..................... 72 |
| Cauliflower/Pea Salad .................. 143 | Green Beans and Sausage .............. 124 |
| Cheese Soup ......................... 145 | Grilled Tomatoes ...................... 76 |
| Cheesy Vegetable Soup ................ 136 | Goodie Casserole ..................... 89 |
| Cinnamon Pickles ..................... 154 | Half & Half Taters .................... 84 |
| Cold Green Beans Sesame .............. 126 | Hash Brown Casserole ................. 84 |
| Cole Slaw ............................ 141 | Hot Broccoli Dip ...................... 110 |
| Company Baked Pork Chop Casserole .... 89 | Hotshot Zucchini ..................... 103 |
| Corn for Freezer ...................... 79 | Knobby Kohlrabi ...................... 133 |
| Corn Zucchini Bake ................... 71 | Mashed Parsnips ...................... 131 |
| Creamed Cucumbers .................. 125 | Mashed Potato Casserole .............. 83 |
| | Melon Balls in Wine ................... 113 |

| | |
|---|---|
| Mixed Mustard Pickles | 136 |
| Mixed Vegetable Casserole | 138 |
| Mom's Rhubarb Pie | 115 |
| Old Country Carrots | 73 |
| One Dish Meal | 139 |
| Oriental Casserole | 134 |
| Oven Caramel Corn | 80 |
| Oven Potatoes | 81 |
| Oven Stew | 139 |
| Overnight Salad | 100 |
| Pan Fried Eggplant | 129 |
| Party Potatoes | 81 |
| Party Potatoes | 83 |
| Party Time Potatoes | 87 |
| Pasta Salad | 91 |
| Peanut Butter Popcorn Balls | 79 |
| Pickled Macaroni | 92 |
| Pineapple Sweet Potato Bake | 112 |
| Potato Chip Dip | 99 |
| Potato Pancakes | 85 |
| Potato Pancakes | 86 |
| Potato Salad | 80 |
| Potato Salad | 82 |
| Potato Salad | 88 |
| Potatoes for Company | 88 |
| Pumpkin Bars | 150 |
| Pumpkin Bars | 152 |
| Pumpkin Cookies | 129 |
| Pumpkin Roll | 153 |
| Quick Caramel Corn | 77 |
| Quick Vegetables Au Gratin | 144 |
| Raw Vegetable Salad | 141 |
| Red Radish Relish | 135 |
| Refrigerator Slaw | 147 |
| Rice Casserole | 144 |
| Rhubarb Bars | 116 |
| Rhubarb Cake | 118 |
| Rhubarb Crisp | 120 |
| Rhubarb Dream Dessert | 117 |
| Rhubarb Dessert | 119 |
| Rhubarb Dessert | 119 |
| Rhubarb Dessert | 121 |
| Rhubarb Jam | 114 |
| Rhubarb Jam | 116 |
| Rhubarb Pie | 118 |
| Rhubarb Swirl | 117 |
| Rushing Around Potatoes | 85 |
| Sauteed Green Tomatoes | 76 |
| Scalloped Potatoes | 72 |
| Snow Pea Soup | 134 |
| Soup and Sausage Casserole | 122 |
| Sour Cream Eggplant | 128 |
| Sour Cream Potatoes | 82 |
| Spinach Salad | 130 |
| Steak in a Crock Pot | 140 |
| Stromboli | 123 |
| Stuffed Green Peppers | 123 |
| Sue's Carrot Cake | 74 |
| Super Carrots | 74 |
| Super Salad | 137 |
| Swede Potatoes | 132 |
| Sweet Dressing for Lettuce | 102 |
| Sweet Pickles | 97 |
| Sweet Potato on the Cheap | 112 |
| Sweet Potato Pie | 110 |
| Taco Salad | 137 |
| Thousand Island Dip for Vegetables | 135 |
| Three Day Pickles | 97 |
| Tomato Preserves | 75 |
| Turnips Turnips Turnips | 132 |
| Vegetable & Chip Dip | 73 |
| Vegetable Combo | 147 |
| Vegetable Dill Dip | 99 |
| Vegetable Dip | 68 |
| Vegetable Dip | 140 |
| Vegetable Pizza | 102 |
| Vegetable Spread | 78 |
| Watermelon Balls | 114 |
| Watermelon Tipsy | 113 |
| You Name It | 138 |
| Yam Crispy Chips | 109 |
| Zucchini Bread | 106 |
| Zucchini Brownies | 106 |
| Zucchini Casserole | 103 |
| Zucchini Chocolate Cake | 104 |
| Zucchini Dip | 103 |
| Zucchini Jam | 105 |

# Carrot Coleslaw

2 C. shredded cabbage
½ C. shredded carrot
1 T. minced onion

1/8 tsp. salt
⅓ C. mayonnaise

In bowl combine cabbage, carrot, and onion. Sprinkle with salt. Add mayonnaise and toss. Cover and refrigerate. Toss again just before serving. Serves 4.

# Carrot Salad, Congealed

1 pkg. lemon or orange gelatin
1 (8 oz.) can crushed pineapple
1½ C. grated carrots

½ C. nuts, chopped
Optional: 1 T. chopped green pepper

Drain pineapple pressing out as much juice as possible. Add water to juice to make 2 C. of liquid. Dip out and discard 1 T. of liquid. Heat liquid to boiling; add gelatin and stir to dissolve. Chill until starting to congeal. Stir in pineapple, carrots, nuts, and green pepper if using. Pour into 1½-quart mold. Chill until firm.

## VEGETABLE DIP

2 C. real mayonnaise
2 C. sour cream
2 T. onion flakes
2 T. parsley flakes

2 tsp. dill weed
2 tsp. Beau Monde
or seasoned salt

Combine all and chill.

## Crock Pot Chicken and Vegetables

1 fryer, cut-up
2 C. water
½ tsp. salt
4 carrots, cut chunk-size
3 small onions, peeled

If counting calories, chicken can be skinned before cooking. Place all ingredients in a slow-cooker. Cook on low heat for 4 to 5 hours until meat pulls away from the bones. Remove chicken and vegetables from the broth to a hot platter. This makes about 2½ cups of chopped chicken and 2 to 3 cups of broth.

## CARROT CAKE

2 C. sugar  
1½ C. oil  
4 eggs  
2 C. flour  
1 tsp. salt  

2 tsp. baking soda  
1 tsp. cinnamon  
3 C. carrots (grated fine)  
½ C. nuts  

FROSTING:  
1 lb. powdered sugar  
8 oz. cream cheese  
¼ lb. butter  

2 T. milk  
2 tsp. vanilla  

Beat sugar and oil well. Add eggs and beat well. Add dry ingredients which have been sifted together. Add carrots and nuts. Bake 1 hour at 300° in 9x13-inch greased pan. Cool.

For Frosting: Beat cream cheese and butter. Add powdered sugar, milk, and vanilla. Spread on cake.

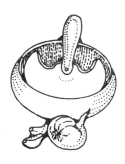

## CARROT LOAF

2 C. grated carrots  
1 C. cracker crumbs  
2 eggs (beaten)  
1½ C. milk  

1 T. chopped onion  
1 tsp. salt  
1 tsp. pepper  
1 T. melted butter  

Mix all together and place in greased casserole. Bake at 350° for 1 hour.

# Anise Carrots

1 tsp. butter
4 C. thinly sliced carrots
3 T. water
¼ tsp. anise seeds, crushed
Salt

Melt butter in skillet; add all ingredients. Stir-fry until carrots are coated.

Cover and cook until crisp-tender. It may be necessary to add water.

Check frequently as carrots burn easily.

## CARROT CAKE

BEAT TOGETHER:
1½ C. oil                    2 C. sugar
3 eggs

SIFT AND ADD:
2 C. flour                   2 tsp. soda
2 tsp. cinnamon              1 tsp. salt

THEN ADD:
2 C. grated carrots (well drained)   1 C. coconut
1 C. nuts                            1 C. crushed pineapple and juice

Bake in a 9x13-inch pan for 1 hour at 350°.

## Corn Zucchini Bake

1 lb. (about 3 to 4) medium zucchini
¼ C. chopped onion
1 T. butter
2 beaten eggs
1 (10 oz.) pkg. frozen corn or 2 C. fresh corn (cooked & drained)
1 C. shredded Swiss cheese
¼ tsp. salt
¼ C. fine bread crumbs
2 T. Parmesan cheese (grated)
1 T. butter (melted)

Slice zucchini 1-inch thick and cook till tender, drain and mash. Cook onion in butter, till tender. Combine beaten eggs, zucchini, onion, corn, and Swiss cheese, and salt; put into a 1-quart casserole. Sprinkle bread crumbs and cheese on top, then pour on butter. Bake for 25 to 30 minutes or until knife comes out clean. Makes 6 servings.

## Carrot, Zucchini Bars

1½ C. flour
¾ C. brown sugar
1 tsp. baking powder
½ tsp. baking soda
½ tsp. ginger
2 eggs (slightly beaten)
1½ C. shredded carrot
1 C. shredded zucchini
½ C. raisins (I like white)
½ C. chopped walnuts
½ C. cooking oil
¼ C. honey
1 tsp. vanilla
1 recipe Citrus cream cheese frosting

Sift or stir together flour, baking powder, baking soda, and ginger. Stir in the brown sugar. Mix eggs, carrot, zucchini, nuts, oil, honey, and vanilla together. Stir flour mixture into the eggs, etc. Spread into a jelly roll pan and bake at 350° for about 20 to 25 minutes (until done). Cool and frost.

CITRUS CREAM CHEESE FROSTING:
2 (3 oz. ea.) pkgs. cream cheese
½ C. butter or margarine
1 tsp. vanilla (may be omitted)
2 T. lemon or orange juice
1 T. lemon or orange peel
4 C. plus powdered sugar

Mix ingredients together until smooth and of spreading consistency; frost bars. Coconut may be sprinkled over top. Makes 36 bars.

## SCALLOPED POTATOES

5-6 medium potatoes (peeled and sliced)
½ medium chopped onion
1 C. shredded cheddar cheese
1 C. sour cream
1 can cream of celery soup
½-1 C. milk
Salt and pepper

Mix sour cream, soup, and milk in a separate bowl. Layer potatoes, soup mixture, onions, and cheese in 1½-qt. greased baking dish. Salt and pepper to taste. Bake at 375° for 1½-2 hours.

## GREEN BEAN BAKE

1 can cream of mushroom soup
1 tsp. soy sauce
3 C green beans (drained)
Dash pepper
1 (3½ oz.) can French fried onion rings

In 1-qt. casserole, stir soup, soy sauce, and pepper until smooth. Mix in beans and ½ can onions. Bake at 350° for 20 minutes. Top with remaining onions and bake 5 minutes more.

## Old Country Carrots

2½ C. sliced raw carrots
½ C. butter
½ C. milk

3 eggs, beaten
2 C. grated Cheddar cheese

Boil carrots until tender. Mash, add butter, milk, and eggs. Fold in Cheddar cheese. Bake at 350° for 40 to 50 minutes.

**VEGETABLE AND CHIP DIP**

1 qt. Hellman's mayonnaise
4-oz. Kraft French dressing
1½-oz. parmesan cheese

1 pkg. Good Seasonings Italian dressing mix

Mix together.

## Super Carrots

5 C. carrots (sliced)
¼ C. water
1½ C. celery (chopped)
1 medium onion (chopped)
1 green pepper (chopped)
1 can tomato soup

1 C. confectioners' sugar
¾ C. vinegar
½ C. salad oil
1 tsp. prepared mustard
1 tsp. Worcestershire sauce
½ tsp. salt

Place carrots and water in a 3-quart casserole. Microwave, covered on full power for 6 to 8 minutes or until barely cooked. Stir halfway through cooking time. Drain and cool. In a large bowl, mix together remaining ingredients. Pour marinate mixture over carrots and refrigerate for several hours.

## Sue's Carrot Cake

1 C. sugar
1 C. flour
1¼ tsp. cinnamon
1 tsp. baking soda
1 tsp. baking powder
½ tsp. salt
¼ tsp. ginger

¼ tsp. ground cloves
½ C. vegetable oil
2 eggs
1½ C. grated carrots
1 (8 oz.) can crushed pineapple
  (drained)

Blend together all dry ingredients in a large mixing bowl. Stir in oil and add eggs, 1 at a time, mixing well after each addition. Blend in carrots and pineapple. Pour batter into greased 8 x 8 x 2-inch dish. Cook in microwave on full power for 7 or 9 minutes or until top springs back; cool.

CREAM CHEESE FROSTING:
1 (3 oz.) pkg. cream cheese
¼ C. margarine

1 tsp. vanilla
2 C. powdered sugar

Beat together cream cheese, butter, and vanilla. Gradually add powdered sugar, beating until smooth. If too thick add a few drops, as desired.

# Garden Tomato Salad

6 tomatoes, unpeeled & chopped
1 red onion, sliced
2 small cucumbers, thinly sliced
6 spinach leaves
¼ C. Italian dressing

Toss all ingredients together. Refrigerate until ready to use. Makes 6 servings.

## TOMATO PRESERVES

6 C. ripe tomatoes (peeled and diced)
6 C. sugar
1 lemon (sliced)
3-4 sticks cinnamon

Cook all over low heat until thick (approximately 40-45 minutes). Put in hot sterilized jars and seal.

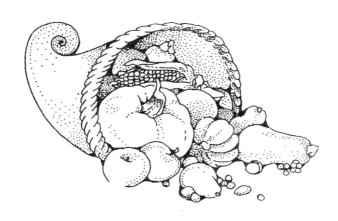

# Sauteed Green Tomatoes

½ C. flour
5 medium green tomatoes,
  sliced ½-inch thick

5 T. bacon drippings
  or cooking oil
Black pepper

Flour both sides of tomatoes and fry in hot drippings in oil until browned and tender. Pepper liberally. Really delicious with meats.

# Grilled Tomatoes

6 firm tomatoes
4 T. butter, melted

Dried basil or curry powder
Salt

Cut tomatoes in half crosswise. Place in greased baking dish. Brush with melted butter and season with salt. Sprinkle lightly with basil or curry powder or a combination of both. Place about 6 inches from broiler heat. Tomatoes should cook slowly. Do not turn. Place in aluminum foil to cook on grill. It will take 10 to 12 minutes.

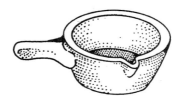

## CARAMEL CORN

1 C. brown sugar  ½ C. margarine
10 marshmallows  2 tsp. white syrup

Heat together; do not boil. Pour over popcorn and mix together.

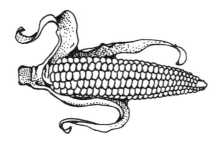

## QUICK CARAMEL CORN

¾ C. brown sugar  ⅓ C. honey
2 T. margarine

Put ingredients in saucepan on low heat until melted. Cook until bubbly. Pour over 10 C. popped corn. Stir until all is coated. Put in shallow pan. Bake for 15 minutes at 350°, stirring once. Dump onto wax paper to cool. Store in covered container.

# Vegetable Spread

¾ C. grated carrots
¼ C. chopped onions
¼ C. chopped radishes
¼ C. chopped celery

¼ C. chopped green pepper
1 T. lemon juice
1 (8 oz.) pkg. cream cheese
(softened)

Sprinkle vegetables with lemon juice. Add to cream cheese and mix well; chill. Serve on or with any crackers.

## FREEZER CORN

35 ears (18 cups)
1 lb. butter

1 pint Half and Half

Cut off raw corn; put in roaster. Add butter and Half and Half. Put in oven at 325°-350° for 1 hour. Stir occasionally so it doesn't burn. Set pan in ice water to cool, then pack and freeze.

## CORN FOR FREEZER

8 C. corn (cut from cob)  
1½ C. water  

¼ C. sugar  
2 tsp. salt  

Bring to a boil (I use an electric fry pan to do this). Boil for 8 minutes. Spread on cookie sheets to cool. When totally cool, fill freezer bags and freeze.

## Peanut Butter Popcorn Balls

*1 C. raw popcorn*  
*1 C. white sugar*  
*1 C. light syrup*  

*1 C. chunky peanut butter*  
*1 tsp. vanilla*  
*1 C. Spanish salted peanuts (optional)*  

Pop corn and keep warm in 200° oven. Bring sugar and syrup to a rolling boil, stirring constantly. Remove from heat. Add peanut butter and vanilla; mix well. Pour over popcorn and mix well. Form into small balls. Optional: Add 1 C. peanuts to sryup mixture before pouring over popcorn.

# Oven Caramel Corn

8 or 9 qt. popped corn  
2 C. brown sugar  
1 C. margarine  
1 tsp. salt  

½ C. white syrup  
1 tsp. burnt sugar or vanilla  
½ tsp. soda  

Mix all ingredients together, except corn and soda; boil for 5 minutes, mixing well and stirring. Remove from heat and add soda; stir in quickly and pour over popped corn, mixing well. Put in 2 large flat pans and place in 250° for 1 hour, stirring 2 or 3 times. Store in tightly closed container.

# Potato Salad

6 potatoes  
3 T. chopped onion  
3 hard-cooked eggs, chopped  

Salt  
1 C. mayonnaise (or more)  

Peel and dice potatoes; boil for about 10 minutes or until tender. Drain and sprinkle potatoes, eggs, and onions lightly with salt. Toss with mayonnaise and chill.

When I open a can of green beans or peas, I always freeze the juice to use when boiling potatoes for salad. It makes the salad zesty and adds extra vitamins.

## Party Time Potatoes

8 or 9 potatoes  
½ pt. sour cream (1 C.)  
1 (8 oz.) cream cheese  

1 tsp. garlic salt  
1 tsp. onion salt  

Cook potatoes; add salt and mash, then add other ingredients. Put in buttered casserole; dot with butter and sprinkle with paprika. Bake, uncovered at 350° for 30 minutes. Cheese may be added on top. These can be stored in refrigerator overnight and heated when desired to use.

## Oven Potatoes

8-10 cooked potatoes  
1 (8 oz.) pkg. cream cheese  
1 (8 oz.) carton sour cream  
Garlic salt, to taste  

Chives  
Butter  
Paprika  

Cream the cheese; add the potatoes and beat until fluffy. Add the sour cream. Stir in the garlic and the chives. Pour into greased casserole and top with butter and paprika. (This recipe is flexible; add more of any ingredient to suit.) It can also be made 2 days ahead and refrigerated.) Before serving, heat in 350° oven for 30 minutes.

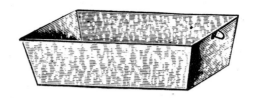

## SOUR CREAM POTATOES

8-10 potatoes
8 oz. cream cheese
1 C. sour cream
Butter

Paprika
Chopped chives
Garlic salt to taste

Cook, drain and peel potatoes. Beat cream cheese and sour cream until well blended. Add the ptoatoes, a few at a time, beating until light and fluffy. Season as desired. Spoon potatoes into greased 2-qt. casserole and dot with butter. Sprinkle with paprika. Bake at 350° for 30 minutes.

## POTATO SALAD

5 lbs. potatoes (boiled and cooled, peeled and sliced)
1 pint Miracle Whip
Milk
Salt and pepper

¼ C. sugar
1½ T. mustard
6 hard boiled eggs (diced)
Onion (chopped)

Place prepared potatoes and onion in large bowl. In separate bowl mix Miracle Whip, enough milk to make creamy, sugar, mustard, salt and pepper. Mix to suit taste. Pour over potatoes and mix well. Fold in diced eggs.

## PARTY POTATOES

8-10 medium potatoes
8 oz. cream cheese
8 oz. sour cream
A little milk
Salt and pepper to taste
Parsley flakes (as desired)
Garlic salt (to taste)
2-4 T. butter

Boil potatoes until tender. Mash potatoes and add cream cheese, sour cream, a little milk, garlic salt, parsley flakes, salt and pepper. Beat until fluffy. Spoon into 2-qt. greased casserole. Top with butter (cut in small slices). Bake at 350° for 30 minutes or microwave until hot. Can be made the day before and refrigerated (also freezes well).

## MASHED POTATO CASSEROLE

5 lbs. or 9 large potatoes
1 (8 oz.) pkg. cream cheese
1 C. sour cream
2 tsp. onion salt
¼ tsp. pepper
1 tsp. salt
Sharp cheddar cheese (shredded)

Peel potatoes and boil; mash with electric mixer but use no milk or butter. Mix in other ingredients except butter and cheese. Place in buttered 2-qt. casserole or 9x13-inch pan. Top with 2 T. butter. Bake for 30 minutes in a 350° oven. Garnish with grated cheese when ready to serve. Serves 10. Can be made a day ahead and refrigerated. Bake 45 minutes if refrigerated.

## DAIRY BANQUET POTATOES

8 or 9 medium potatoes  
1 stick butter  
1 pt. Half and Half  
1 tsp. salt  
¼ lb. cheddar cheese

Cook potatoes with skins on. Cool, peel, slice, and place in a 9x13-inch pan. Melt butter, add Half and Half and salt. Pour over potatoes. Grate cheese over potatoes. Bake at 350° for 1 hour.

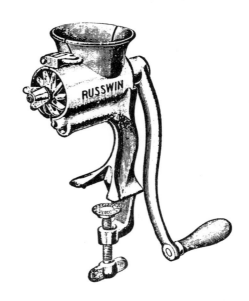

## HASH BROWN CASSEROLE

2 lb. bag Ore-Ida hash browns  
Salt and pepper to taste  
Grated onions  
Cheddar cheese (grated)  
½ C. oleo  
3 oz. cream cheese  
1 can cream of mushroom soup  
½ soup can of milk (5 oz.)

Place hash browns and onions in greased 9x13-inch pan. Season with salt and pepper. Combine cream cheese, oleo, mushroom soup and milk. Pour over potatoes. Cover with grated cheese. Cover and bake at 350° for 1½ hours.

## Rushing Around Potatoes

8 to 10 potatoes
1 onion

1 stick margarine
Salt & pepper

Peel and slice potatoes and onion. Place in large glass cake pan. Salt and pepper. Slice margarine over the top. Cover with Saran Wrap and microwave on high for 10 to 12 minutes. Turn dish and microwave another 10 minutes. Let stand for 5 to 10 minutes before serving. Can melt cheese slices or Cheez Whiz on top while standing.

## Potato Pancakes

2 medium potatoes, shredded
1 carrot, grated
1 onion, minced

⅓ C. flour
2 eggs

Combine vegetables. Stir in flour, add beaten egg. Cook on lightly greased griddle or skillet for about 5 minutes on one side; turn and flatten pancake with spatula. Cook an additional 5 to 6 minutes.

## Beat-Up 'Taters

8 to 10 potatoes (peeled)  
½ C. water  
1 (8 oz.) pkg. cream cheese  
1 carton French onion soup  
½ C. milk  
1½ tsp. salt  
1/8 tsp. pepper  
Garlic salt (optional)  
2 T. margarine  
Paprika  

Halve and quarter potatoes. Combine potatoes and water in a 3-quart glass casserole. Microwave on high, covered for 20 minutes; drain. Blend together cream cheese and onion dip in a large mixing bowl with electric beater. Add hot potatoes, several pieces at a time. Add, beating until light and fluffy. Stir in salt, pepper, and garlic salt. Spoon into 3-quart casserole. Dot with margarine. Microwave on high, covered for 10 minutes, stirring halfway through microwave time. Sprinkle with paprika.

## Potato Pancakes

2 eggs (beaten)  
¼ C. skim or 2% milk  
2 C. grated raw potatoes (drained)  
1 T. grated onion  
3 T. flour  
1 tsp. salt  
Dash pepper  

Combine beaten eggs and milk in mixing bowl. Add grated potatoes that have been drained well. Add onion, flour, salt, and pepper; mix well. Drop the mixture by tablespoonfuls onto a hot, lightly greased griddle. Stir the mixture before dropping each tablespoonful. Cook the cakes until they are well browned and crisp on the bottom, then turn and cook on the other side. Cakes must cook slowly or they will become scorched before cooking through. Makes 1-1½ dozen 3-inch pancakes.

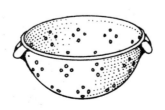

## Party Potatoes

10 large potatoes
8 oz. sour cream
8 oz. cream cheese

Salt and pepper
Garlic salt (optional)

Boil potatoes until done. Mash thoroughly. Add sour cream and cream cheese, salt, and pepper to taste. Garlic is optional. Beat all together thoroughly. Put into greased casserole. This can be frozen and reheated and still tastes fresh.

## Potatoes for Company

12 large potatoes  
¼ C. margarine  
1 (8 oz.) pkg. cream cheese  
1 (8 oz.) carton sour cream  
1 tsp. chives or onion

Mash potatoes and add the next 4 ingredients. Put in casserole or 9×13-inch pan, dot with butter. Bake at 350° for 45 minutes. (You can make these ahead of time.)

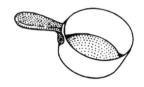

## Potato Salad

5 C. potatoes (cooked, cubed, or sliced)  
2 tsp. sugar  
2 tsp. vinegar  
½ C. onions (chopped)  
1½ tsp. salt  
1½ C. Miracle Whip  
4 hard cooked eggs  
½ C. celery if desired

Combine, chill and serve.

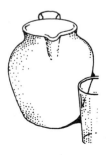

# Spuds

1 dozen VERY small new potatoes
¼ C. sour cream or yogurt
1 T. grated carrot

1 T. chopped green onion & tops
4 slices bacon, crisp & crumbled

Boil the potatoes until tender. When cool, peel and slice into halves. Using a teaspoon, hollow out centers. Stir together sour cream, carrot, and onion. Fill potatoes with this mixture and top with crumbled bacon. Serves 3 to 4.

## Goodie Casserole

4 pork chops
4 medium potatoes (sliced)
1 can cream of mushroom soup

2 medium onions (sliced)
Velveeta cheese

Place chops in pan; salt and pepper. Cover with potato slices; then sliced onions, a layer of cheese; spread with a can of soup and another layer of cheese. Cover and bake at 350° for 1 hour.

# Garlic Potatoes

8 or 10 potatoes (mashed)  
8 ozs. cream cheese  
1 C. sour cream  
½ tsp. garlic salt  
½ tsp. onion powder  

Mash your potatoes in mixer. Add cream cheese and sour cream. Add seasoning to taste (the garlic and onion) and salt. Brush with butter and bake at 350° for 30 minutes. Bake in round casserole.

## BROWNED PAPRIKA POTATOES

6 medium-sized potatoes  
1 T. melted bacon drippings or other fat  
¼ C. fine corn flake crumbs  
1 tsp. paprika  
1 tsp. salt  

Pare potatoes and boil in salted water until about half done. Brush with melted fat. Mix corn flake crumbs with paprika and salt. Roll potatoes in this mixture until well coated. Bake in shallow, greased, baking pan in 425° oven about 45 minutes. Serves 6.

## PASTA SALAD

2 C. cooked pasta (twists)  
2 hard boiled eggs  
¾ C. chopped green pepper  
¾ C. red onion  
2 T. parsley (fresh or dried)  

MIX THESE INGREDIENTS IN A JAR AND SHAKE WELL:  
⅓ C. vegetable oil  
¼ C. wine vinegar  
1 tsp. salt  
½ tsp. pepper  
¼ tsp. garlic powder  
¼ tsp. oregano  

Put pasta (twists) in bowl. Add chopped eggs, green pepper, chopped red onion, and parsley. Pour the jar of dressing over above ingredients and toss. Can add any one of the following: pieces of tomato or pepperoni, green or black olives. A seafood pasta with shrimp or Sea Legs.

## GARDEN SALAD

1 box Kraft macaroni and cheese  
1 C. diced, peeled tomatoes  
1 C. diced unpeeled cucumber  
½ C. sour cream  
¼ C. Miracle Whip  
¼ C. chopped radishes  
¼ tsp. salt  

Prepare macaroni and cheese as directed on package. Add remaining ingredients. Mix lightly and chill.

## PICKLED MACARONI

½ lb. rigatoni macaroni
1 T. salad oil
1½ C. white vinegar
¾ C. sugar
½ tsp. salt and pepper
½ tsp. Accent
½ tsp. garlic powder
1 T. parsley flakes
1 T. dry mustard
1 small onion (chopped)
1 medium cucumber (chopped)

Cook macaroni until tender - not soft - drain well. Coat with salad oil; add macaroni to brine made from remaining ingredients. Chill well and serve.


## GARDEN MACARONI SALAD

2½ lbs. macaroni (big shells)
½ lb. radishes
2½ cucumbers
½ stalk celery
1 green pepper
½ lb. carrots
1/8 C. dehydrated onions

DRESSING:
½ C. vinegar
1 C. sugar

Salt and pepper to taste
1 qt. (2 bottles) Kraft Cole Slaw dressing

Pour over cooked macaroni and add the rest of the above mixture.

# CUCUMBER SALAD

3-4 cucumbers (6-inches or 7-inches long)

3-4 onions (sliced thin, red if available)

DRESSING:
1½ C. Miracle Whip
1½ C. sugar
2 tsp. mustard

1 tsp. horseradish
2 T. vinegar
Half and Half as needed

Peel cucumbers and cut ends off to start of seeds and slice very thin. Mix dressing ingredients together except Half and Half. Mix with spoon or egg beater (egg beater is best). Add Half and Half a little at a time so dressing won't get too thin. Add dressing to cucumbers and onions; mix well. Store in refrigerator. This is better made a day ahead. No salt - will make it watery.

# CUCUMBER AND TOMATO VINAIGRETTE

2 cucumbers (pared and sliced)
1 red onion (sliced)

2 medium ripe tomatoes (sliced)
½ C. bottled vinaigrette salad dressing

Combine cucumber, tomato, onion slices and dressing in a salad bowl; toss until thoroughly coated. Chill until ready to serve. Toss again before serving.

## CUCUMBER SALAD

2 large or 3 small cucumbers (sliced thin)
2 medium purple onions (sliced thiin)

DRESSING:
1½ T. vinegar
¾ C. sugar
1 tsp. (heaping) mustard

Salt and pepper or seasoned salt and seasoned pepper

¾ C. Miracle Whip
Half and Half as needed

Place sliced cucumbers and onions (separate onions into single rings) in bowl. Salt and pepper them, toss lightly. Blend together dressing ingredients and add Half and Half to thin (about half of a small carton). Beat well with egg beater, pour over cukes and onions, stir well. Cover and refrigerate for several hours, overnight is best. Serve over a thick slice of tomato. It is good without the tomato too.

## CUCUMBER SALAD

3-4 cucumbers (peeled and sliced thin)
3 onions (sliced thin)
2 C. mayonnaise or Miracle Whip

1 C. sugar
1 T. mustard
1½ T. vinegar
1 T. horseradish

Mix mayonnaise, sugar, mustard, vinegar, and horseradish; blend well. Pour over sliced cucumbers and onions, stir lightly until they are coated well.

## CUCUMBER SNACKS

1 cucumber (sliced thin)  
8 oz. cream cheese  
1 loaf party rye bread  
1 pkg. Good Season's dry Italian dressing

Mix cream cheese and dry dressing. Spread on rye bread slices. Top with slice of cucumber and sprinkle with dill weed. Chill.

## CUCUMBER SALAD

2 cucumbers (peeled and sliced)  
2 small onions (sliced)

½ C. sugar  
1 tsp. salt  
¼ C. vinegar

Place cucumbers and onions in a bowl. Bring rest of ingredients to a boil and then pour over cucumbers and onions. Cover and refrigerate overnight.

## Easy Dill Pickles

9 C. water  
1 C. salt  
2 C. vinegar  
Dill

Combine ingredients. Put on stove and heat until bubbly. Do not boil. Put plenty of dill in jar, then pour over baby cucumbers. Put seal on, let set for couple of weeks - ready to eat. Boil lid in water before putting on jar.

## CUCUMBERS WITH SOUR CREAM

2 medium cucumbers  
1 tsp. salt  
1 C. sour cream  
2 T. vinegar  
1 T. onion (chopped)  
1 T. sugar  
¼ tsp. salt  
1/8 tsp. pepper

Pare and slice the cucumbers and place in a small bowl. Sprinkle with salt and chill for 30 minutes. Combine the sour cream, vinegar, onion, sugar, salt and pepper. Chill. Drain the cucumbers and press excess water by placing the cucumbers in the folds of a clean towel. Add sour cream sauce.

## Sweet Pickles

2 qts. whole dill pickles
3 C. sugar
1 C. cider vinegar

2 C. water
2 tsp. celery seed

Pour off vinegar on pickles and wash dill pickles and cut in chunks and put back in jar. Fill jar with ice cold water and place in refrigerator. Mix sugar, water, and vinegar; boil and simmer until sugar is dissolved. Cool and put in refrigerator until cold. Take pickles out of refrigerator and drain well. Put lid back on and put jar in refrigerator upside down. Drain again and pour 1 tsp. celery seed in each jar. Fill with syrup and place in refrigerator. Let set a few days before using.

## Three Day Pickles

1 C. white vinegar
2 C. white sugar

1 (46 oz.) jar Vlasic Zesty Kosher Dills (or Crunchie dills)

Pour liquid off pickles (discard) and wash pickles in water; slice pickles. Heat vinegar and sugar to a boil; let boil for 1 minute. Pour over pickles in jar. Refrigerate when cool. Next day, pour off vinegar solution and boil it again and pour back over pickles. Do this again on the third day. Now they are ready to eat.

# Cabbage Salad

1 medium head cabbage (shredded)
1 large carrot (shredded)
½ C. diced cucumber
¼ C. chopped green pepper
¼ C. chopped onion
½ C. diced celery
3-4 radishes (cut up)

Combine all of the above.

DRESSING:
1 C. Miracle Whip
½ C. garlic French dressing
1 T. sugar
1 tsp. salt
Pepper to taste
1 T. vinegar

Mix dressing ingredients well. Pour over vegetables.


## GARDEN PATCH DIP

8-oz. cream cheese
2½ T. sour cream
2 T. minced onion
2 T. minced radish
2 T. horseradish
1 C. grated carrot
¼ tsp. salt
3 drops lemon juice

Beat the cream cheese with the sour cream. Stir in the onion, radish and horseradish. Add the carrots, salt, and lemon juice.

## POTATO CHIP DIP

1 can tomato soup
1 box lemon Jell-o
1 (8-oz.) pkg. cream cheese
1 C. celery (chopped fine)

1 C. Miracle Whip
3 bunches onions
Dash of Worcestershire sauce

Heat tomato soup, add Jell-o and stir until dissolved, cool. Add the Miracle Whip and cream cheese, beat with the mixer. Add the green onions using some tops chopped up. Add the celery and a dash of Worcestershire sauce. Make ahead and let the flavor blend.

## Vegetable Dill Dip

1 C. Miracle Whip
1 C. sour cream
1 pkg. dry onion soup
1 T. Worcestershire sauce

1 T. dill weed
1 T. shredded parsley flakes
3 drops Tabasco sauce

Stir together all ingredients and store in refrigerator.

## OVERNIGHT SALAD

1 head lettuce (washed and drained well)
1 small red onion
1 C. Hellmann's real mayonnaise
1 small can Parmesan cheese
½ head cauliflower (washed and dried and cut into bits)
1 lb. bacon (fried crisp)
¼ C. sugar

All ingredients should be dry or salad will be soupy. Cut onion into rings or chop. Layer in serving bowl and make dressing but do not mix until ready to serve. To make dressing, just blend all 3 ingredients well. This may be made the night before and stored in refrigerator in Tupperware or airtight container, but Do Not toss until ready to serve.

## Curry Dip For Vegetables

2 C. mayonnaise
2 tsp. curry powder
2 tsp. vinegar
2 tsp. horseradish (opt.)

2 tsp. garlic salt
½ tsp. garlic powder
2 tsp. minced onion

Mix together and serve with vegetables. Makes 1 pint.

## CAULIFLOWER-BROCCOLI SALAD

1 head cauliflower
3 bunches broccoli
Pimento

1 red onion
1 green pepper

Chop the above and place in large bowl.

MIX:
1 C. mayonnaise
½ C. oil
½ C. vinegar

½ C. sugar
Salt and pepper

Pour over above mixture. Chill and serve.

# Vegetable Pizza

1 can (8) Pillsbury crescent rolls
1 (8 oz.) pkg. cream cheese
⅓ C. Miracle Whip
½ tsp. garlic powder
½ tsp. dill weed

Chopped vegetables: broccoli, cauliflower, green pepper, ripe olives
Grated cheese: Colby, Monterey Jack

Press crescent rolls into a pizza pan and bake according to directions on package. Cool. Mix cream cheese, Miracle Whip, garlic powder, and dill weed. Spread over crust. Sprinkle chopped vegetables over cheese mixture and top with grated cheeses.

# Sweet Dressing for Lettuce

2 eggs (well-beaten)
1 tsp. salt
1 tsp. mustard
Pinch of paprika
3 T. flour

1 C. sugar
1 T. vinegar
1 C. water
Lump of butter

Mix all ingredients thoroughly and put in double boiler or heavy pan over low heat. Cook until thick, about 5 minutes.

# Hotshot Zucchini

1 lb. ground beef
1½ C. ground zucchini
¼ C. butter
1½ C. cheese (cut up)

½ tsp. salt in meat
1 C. milk
1½ C. cracker crumbs
3 eggs

Brown meat and drain. Mix with the rest of ingredients. Cover and bake at 325° for 40 minutes.

# ZUCCHINI DIP

2 meduim zucchini (chopped, 2 C.)
½ C. tomato juice
1 T. chopped onion
¼ tsp. salt

1/8 tsp. dried basil (crushed)
1 (8 oz.) pkg. light cream cheese
½ slice bacon (crisp cooked and crumbled)
Assorted vegetables for dipping

In a saucepan, combine the zucchini, tomato juice, onion, salt, and basil. Simmer, covered, for 10 minutes. Turn mixture into blender container. Add the cheese. Cover and blend until mixture is thoroughly combined. Remove from blender. Cover and chill. Just before serving, if desired, sprinkle top with crumbled bacon. Serve with assorted vegetable dippers. Makes 1¾ C. (Per tablespoon: 24 calories, 2 grams fat, 60 milligrams sodium.) Count as ½ fat.

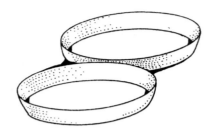

# Zucchini Romano

4 C. sliced zucchini
1 C. onion rings
4 T. butter

1 large tomato, chopped
2 T. Romano cheese, grated

Stir-fry zucchini and onions in butter 8 to 10 minutes. Add tomatoes and cook another 4 minutes. Pour into warm serving dish and sprinkle with Romano cheese.

## ZUCCHINI CHOCOLATE CAKE

½ C. soft butter
½ C. oil
1¾ C. sugar
2 eggs
½ C. milk + 1 T. vinegar
1 tsp. vanilla
2½ C. flour

4 T. cocoa
½ tsp. salt
½ tsp. baking powder
1 tsp. soda
½ tsp. cinnamon
2 C. shredded zucchini
½ C. chocolate chips

Cream oil, oleo, sugar. Add eggs, vanilla, and milk soured with vinegar. Beat well. Stir in dry ingredients and beat well. Stir in zucchini - beat well. Put in a greased and floured 9x13-inch pan. Sprinkle with chocolate chips. Bake at 325° for 40-45 minutes or until it springs back when touched lightly.

## Zucchini Plus

3 C. zucchini, sliced
½ C. onion rings
2 C. carrot sticks

2 T. butter or margarine
Salt

Stir-fry carrots in butter or margarine for 3 minutes; add zucchini, onions, and salt. Continue cooking until vegetables are crisp-tender. Salt, to taste.

## Zucchini Jam

5½ C. grated zucchini
6 C. sugar
1 C. water

2 T. lemon juice
1 (20 oz.) can crushed pineapple
2 (3 oz.) pkgs. Jello (any flavor)

Boil 6 minutes zucchini, sugar, and water. Add lemon juice and pineapple. Boil 6 minutes more. Add two 3 oz. each pkgs. Jello. Boil 6 minutes more. Pour hot mixture into jars, put on lid and screw band. Jars will seal without processing if you put on lids immediately after you pour in hot mixture.

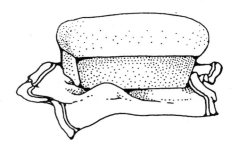

# Zucchini Bread

3 eggs
1 C. cooking oil
2¼ C. sugar
2 C. grated zucchini
1 tsp. vanilla
¼ tsp. black walnut flavoring

1 T. cinnamon
1 tsp. salt
1 tsp. soda
1¼ tsp. baking powder
3 C. flour
1 C. chopped nuts

Beat eggs, add oil, sugar, zucchini, and flavoring; mix well. Sift in dry ingredients. Mix and fold in nuts. Spoon into 3 greased and floured loaf pans, filling ⅔ full. Bake at 350° for 1 hour or until done.

# Zucchini Brownies

CREAM TOGETHER:
1½ C. sugar
2 eggs

½ C. butter

MIX AND ADD:
2 C. flour
½ tsp. cinnamon
1 tsp. soda

2 T. cocoa
2 C. grated zucchini
   (seeds removed)

Stir in 1 C. chocolate chips. Spoon into a greased and floured pan. Makes 25 brownies, 167 calories each. Use 9 × 13-inch pan. Bake at 350° for 45 minutes.

# Dilly Zucchini

2 medium zucchini  
3 T. butter  
Dill weed, fresh or dried  

Optional: red bell pepper strips to garnish

Cut 2 unpeeled zucchini in half lengthwise. Cook, covered in boiling salted water for 12 to 15 minutes or until tender. Drain, brush with melted butter, sprinkle with dill weed, and bell pepper, if using. If you don't have bell pepper, a sprinkle of paprika will add some color.

# Broccoli Sesame

4 C. broccoli florets  
4 tsp. toasted sesame seeds  
¾ tsp. sesame seed oil  

1½ T. lemon juice  
Optional: Chopped red bell pepper for garnish

Simmer broccoli until crisp-tender; drain. Stir together remaining ingredients. Pour over broccoli and garnish with chopped red bell pepper, if desired.

## BROCCOLI CASSEROLE

2 T. onion (chopped)
2 T. celery (chopped)
¼ C. butter
2 C. cooked rice

1 can cream of mushroom soup
8 oz. jar Cheez-Whiz*
10 oz. pkg. chopped broccoli

Cook broccoli; drain. Saute onion and celery in butter. Mix all ingredients together and place in greased casserole. Bake at 350° for 30 minutes.
*Velveeta cheese may be used instead of Cheez-Whiz, but add ¼-½ C. milk.

## BROCCOLI DELIGHT SALAD

1 large bunch fresh broccoli
¼ C. diced onion
1 C. raisins

1 C. sunflower seeds
10 strips bacon (fried and diced)

DRESSING:
½ C. mayonnaise
1 T. vinegar

3-4 T. sugar

Place diced, well drained broccoli in large bowl. Add raisins, onion, sunflower seeds and bacon. Mix dressing ingredients. Pour over salad and toss well. Chill for ½ hour.

## Fresh Broccoli Salad

2 bunches raw broccoli (chopped) or 1 bunch broccoli and 1 bunch cauliflower
1 pint fresh mushrooms (sliced)
6 slices bacon (browned and crumbled)
1 medium onion (opt.)
2 hard boiled eggs (diced)
1 small jar stuffed green olives (chopped, opt.)
Mayonnaise

Mix all ingredients with mayonnaise to desired consistency.

## Yam Crispy Chips

6 yams
Water
Peanut oil, or other for frying

Peel yams and slice on thinly as possible. Soak chips for 4 hours. Drain and pat dry with cloth or paper towels. Heat oil to 400° and cook chips to a golden brown; drain. Sprinkle with salt and enjoy!

## SWEET POTATO PIE
*(Makes one 9-inch pie)*

1¼ C. cooked, mashed sweet potatoes
3 eggs (beaten)
½ C. brown sugar
1 T. honey
1 tsp. mace
1 tsp. salt
1 C. milk
1 (9-inch) unbaked pie crust

Combine all ingredients for filling; blend well. Pour into pie crust. Bake at 425° for 20 minutes. Reduce heat to 350° and bake 25-30 minutes.

## HOT BROCCOLI DIP

½ C. onion (chopped)
½ C. celery (chopped)
½ C. mushrooms (chopped)
1 pkg. (10-oz.) frozen broccoli
1 can mushroom soup (chopped) or about 10-oz. fresh broccoli
1 pkg. (6-oz.) garlic cheese or equivalent cheese seasoned with garlic powder

3 T. butter or margarine

Saute onion and celery in butter until golden. Add mushrooms and sauce until light brown; keep warm. Cook broccoli in unsalted water and drain thoroughly. While hot, stir in cheese and mushroom soup. Serve warm with chips, crackers, or fresh vegetables.

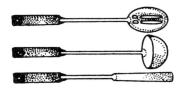

## SWEET POTATO-APPLE CASSEROLE

4 large sweet potatoes		4 large Jonathan apples

SAUCE:
3 T. butter (melted)		1 C. brown sugar
¼ C. condensed milk		2 T. cornstarch
½ C. Karo syrup

Cook sweet potatoes; peel and cut in chunks. Place in greased casserole. Quarter apples, remove seeds and leave peel on. Add to potatoes. Heat and mix in saucepan the butter, condensed milk, Karo syrup. Add brown sugar and cornstarch. Pour over potatoes and apples. Cover and bake at 350° for 1½ hours.

## CARROT CASSEROLE

4 C. sliced carrots		1 small onion (chopped) or
1 C. Velveeta cheese (cubed)	    onion flakes
½ stick butter or margarine	Crushed potato chips

Cook carrots in water for 5 minutes (should be crisp). Place in buttered casserole dish and stir in onion and butter. Sprinkle cheese cubes on top, then crushed potato chips. Bake at 350° for 30 minutes. (I put in a little salt and pepper.)

# Pineapple Sweet Potato Bake

3 medium sweet potatoes  
3 T. melted butter  
1 small can crushed pineapple  
½ tsp. salt

Boil unpeeled sweet potatoes until tender; peel and slice into buttered casserole. Spoon pineapple over sweet potatoes; sprinkle with melted butter and salt. Bake at 350° until light brown and pineapple juice has thickened.

# Sweet Potato on the Cheap

1 butternut squash  
1 C. brown sugar  
¼ C. water  
¼ C. butter  
½ tsp. salt

Pare squash and cut into chunks. Place in buttered casserole. Make syrup of brown sugar, water, butter, and salt. Pour over squash. Cover lightly with aluminum foil. Bake at 325° approximately 1 hour, basting occasionally. Uncover last half of baking time.

## Watermelon Tipsy

2 C. watermelon balls
1 C. lime sherbet
4 lemon slices

Place melon balls in four serving dishes. Top each serving with a scoop of lime sherbert. Garnish with a lemon slice.

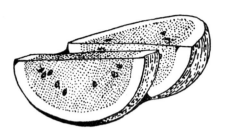

## Melon Balls in Wine

2 (10 oz. ea.) pkgs. frozen mixed melon balls or 4 C. fresh melon balls

3 tsp. sugar
1 C. dry white wine

Completely thaw melon balls. When melon balls are at room temperature; drain well. Place fruit in large bowl; sprinkle with sugar and add wine. Chill in refrigerator, turning occasionally. Serve in compote with toothpicks to spear.

## Watermelon Balls

1 watermelon
1 C. grapefruit juice
⅓ C. sugar
½ C. rum

Cut off top ¼ of the watermelon. Remove all the watermelon meat and make a sawtooth design on the edge of the large part of the melon. Remove seeds from enough melon to puree one cup of juice. Bring grapefruit juice to a boil and add sugar; stir to dissolve and cool. In a bowl combine juices and rum.

Shape the rest of the watermelon into melon balls and return to melon. Cover with juice.

Top melon with plastic wrap. Refrigerate 4 to 6 hours or overnight. Use more rum if you want to, I don't care.

## Rhubarb Jam

2½ lbs. rhubarb, sliced thin
1 C. water
6 C. sugar
1 box Sure-Jell
½ tsp. margarine

Combine rhubarb and water in pan. Simmer, covered for about 12 minutes or until rhubarb is soft.

Measure 4½ cups cooked rhubarb; pour into saucepan. Add 1 box Sure-Jell. Bring to a rolling boil (the boiling cannot be stirred down). Add sugar and bring to a boil, stirring constantly with wooden spoon. Continue boiling and stirring for 1 minute. Remove from heat and skim off foam. Pour immediately into sterilized jars. Wipe tops clean and cover with sterilized lids.

# Mom's Rhubarb Pie

4 C. rhubarb, cut in  
  1-inch pieces  
1½ C. sugar  
6 T. flour  

3 T. butter  
Pastry for 2-crust pie  

Stir together sugar and flour. Sprinkle 4 T. over crust in pan. Sprinkle remaining sugar-flour mixture over rhubarb and stir quickly to coat it. Pour into pie crust and dot with butter. Moisten edge of bottom water. Adjust top crust, press, and trim. Make a few slits in the top so steam can escape.

Mom used to sprinkle a little sugar on the top crust, which makes it brown nicely. Bake at 450° for 15 minutes, turn heat to 350° and bake another 45 minutes.

(To soften ice cream for pie, leave out of the freezer for 10 to 20 minutes.)

# Rhubarb Cake

4 C. rhubarb (diced)  
1 pkg. strawberry Jello  

1 C. sugar  
1 small pkg. Jiffy cake  
1 C. water  

Spread ½ rhubarb in bottom of casserole. Sprinkle with ½ of the Jello, sugar, and dry cake mix. Repeat, topping with dry cake mix. Pour the water very slowly over all. Bake at 350° or until nicely browned.

## RHUBARB BARS

Combine in medium saucepan. Bring to a boil. Cook until thick and bubbly, then set aside:

3 C. cut up rhubarb  
1½ C. sugar  
2 T. cornstarch  
¼ C. water  
1 tsp. vanilla  
½ tsp. baking soda  

In large bowl cut together until crumbly:  
1 C. butter  
½ C. walnuts  
1½ C. oatmeal  
1½ C. flour  
1 C. brown sugar  

Pat ¾ of crumb mixture in 9x13-inch pan. Spread rhubarb over crust. Sprinkle rest of crumbs over top. Bake at 375° for 30-35 minutes.

## RHUBARB JAM

5 C. diced rhubarb  
3 oz. Jello (red)  
3 C. sugar  

Makes 2 pints. Cook rhubarb and sugar until mushy. Put in blender to get out large chunks. Put back in pan and heat and add Jello and stir until dissolved. Put in hot jars. Can be frozen in jars, does not have to be canning jars, just a jar with a lid.

## RHUBARB DREAM DESSERT

1 C. flour
5 T. powdered sugar
½ C. soft butter
3 beaten eggs
2 C. sugar

¼ C. flour
¾ tsp. salt
4 C. finely chopped rhubarb
Cool Whip

Blend first 3 ingredients. Press in 9x13-inch pan. Bake at 325° for 10 minutes. Beat eggs, sugar, flour, and stir. Add rhubarb and pour over crust. Bake 35 minutes at 350°. Cool. Serve with Cool Whip.

## RHUBARB SWIRL

CRUST:*
2 C. flour
1½ sticks soft butter

4 T. brown sugar

FILLING:
3 C. diced rhubarb
3 oz. red Jello
1 small pkg. instant vanilla pudding

1½ C. milk
8 oz. Cool Whip
¾ C. sugar

Pour ¾ C. sugar over rhubarb. Let set 1 hour. Combine crust ingredients and press in 9x13-inch pan. Bake at 350° for 10 minutes. Cool. Simmer rhubarb/sugar mixture until tender about 8-10 minutes. Add Jello and set aside until syrupy. Mix pudding and milk. Fold in Cool Whip. Mix well. Pour syrupy rhubarb mixture into pudding mixture and swirl. Pour onto cooled crust. Chill several hours or overnight. *May also make graham cracker crust instead of above.

## RHUBARB PIE

2 C. rhubarb (sliced thin)  
4 T. flour  
2 eggs  

2 C. sugar  
1 T. vanilla  

Mix flour, sugar, rhubarb; add eggs, then add vanilla. Beat until not grainy. Bake in one crust. Bake at 350° about 1 hour.

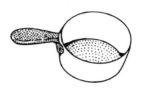

## RHUBARB CAKE

1½ C. brown or white sugar  
½ C. shortening  
1 egg  
1 tsp. vanilla  
1 tsp. soda  

1 C. sour cream or buttermilk  
2 C. flour  
2 C. ½-inch cut rhubarb  
½ C. sugar  
1 tsp. cinnamon  

Cream together sugar, shortening, egg and vanilla. Add soda, buttermilk (or sour cream) and flour. Fold in rhubarb. Combine sugar and cinnamon; sprinkle over top of cake. Bake in a 9x13-inch pan at 350° for 45 minutes.

## Rhubarb Dessert

Rhubarb  
1½ C. sugar  
2 (3 oz.) boxes Jello (strawberry, raspberry or cherry)  
1½ C. flour  
½ C. powdered sugar  
¾ C. butter

In a 9 x 13-inch pan, cut up rhubarb to fill the pan ¾ full. Mix the sugar and Jello together and mix in rhubarb well. Mix the flour, powdered sugar and butter until crumbly, then sprinkle on top of the rhubarb mixture, evenly. Bake at 350° for 30 minutes. Serve with Cool Whip or other whipped topping.

## Rhubarb Dessert

4 C. rhubarb (cut up)  
1 (3 oz.) pkg. strawberry Jello  
¾ C. sugar  
1 pkg. yellow cake mix  
½ C. melted butter  
1 C. cold water

Grease 9 x 13-inch pan, arrange rhubarb on bottom. Sprinkle dry Jello over rhubarb, pour sugar over Jello, then the dry cake mix. Drizzle melted butter over top, followed by cold water. Bake at 350° for 45 minutes. Serve with Cool Whip or ice cream.

# Rhubarb Torte

3 C. diced rhubarb
1 C. sifted flour
1 C. sugar
¾ T. salt

1 unbeaten egg
½ C. shortening
1 tsp. baking powder

Place rhubarb in square pan. Put ½ C. sugar over rhubarb. Mix flour, ½ C. sugar, salt, egg, shortening, and baking powder; sprinkle over rhubarb. Bake in moderate oven.

# Rhubarb Crisp

4 C. cut rhubarb
1¼ C. sugar
½ tsp. cinnamon
1 T. flour
¾ C. flour

½ C. oatmeal
½ C. packed brown sugar
½ C. margarine
Salt

Place rhubarb in shallow pan. Combine sugar, flour, and cinnamon. Sprinkle over fruit. For Topping: Mix ¾ C. flour, brown sugar, and a dash of salt. Cut in margarine and stir in oatmeal. Sprinkle over fruit and bake at 350° for 1 hour. Makes 8 servings. 345 calories.

## Rhubarb Dessert

1½ C. flour  
7 T. powdered sugar  
¾ C. oleo  
2½ C. sugar  
¼ C. flour  

1 tsp. vanilla  
3 beaten eggs  
Dash of salt  
1 C. evaporated milk  

Crumb flour, powdered sugar, and oleo; pat in ungreased 9 x 13-inch pan. Bake at 325° for 15 minutes. Sprinkle 4 C. or more rhubarb on hot crust. Then pour over the rest of the ingredients. Bake at 325° for 45 to 60 minutes.

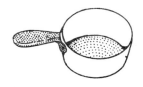

## Belle Pepper Sauce

4 bell peppers  
4 medium onions, sliced  
2 tsp. olive oil  

3 T. bouillon or broth  
Salt  

Clean peppers and cut into strips. Saute' peppers and onions in oil for 6 to 8 minutes. Add bouillon or broth and cook and stir another 2 to 3 minutes. Salt, to taste. Nice served over plain rice or noodles. Serves 4.

## SOUP AND SAUSAGE CASSEROLE
*(Quite a change for a crowd-conscious casserole - a colorful and tasty rice mixture paired with delicately browned pork sausage patties.)*

| | |
|---|---|
| 6 lbs. pork sausage | 6 (2 oz. ea.) envelopes dry chicken noodle soup mix |
| 4 C. chopped celery | ⅔ C. toasted slivered almonds |
| 4 medium onions (chopped, about 2 C.) | ½ tsp. saffron* |
| 2 C. chopped green pepper | 3 qts. boiling water |
| 4 C. uncooked regular rice | Snipped parsley |

Heat oven to 350°. Shape sausage into 48 patties or, if sausage is in rolls, cut each roll into 8 slices. Brown sausage in large skillet over medium-low heat about 3-4 minutes on each side; remove and drain. Pour fat from skillet, leaving just enough to coat bottom. Add celery, onion, and green pepper; cook and stir until onion is tender. In each of 4 ungreased baking pans, 13x9x2-inches, mix ¼ of the onion mixture (about 1½ C.), 1 C. rice, 1½ envelopes soup mix and ¼ of the almonds. Dissolve saffron in boiling water; stir 3 C. into each pan. Arrange 12 sausage patties on mixture in each pan. Cover; bake about 45 minutes or until rice is tender and liquid is absorbed. Garnish with snipped parsley. (Makes 24 servings, 2 sausage patties and 4½-inch square rice mixture per serving.) *If desired, omit saffron; add 10 drops yellow food color to each pan with the boiling water.

## STROMBOLI

1 lb. ground beef
½ C. diced onion
½ C. diced green pepper
½ tsp. garlic powder

1 (15½ oz.) jar spaghetti sauce
2 (8 oz. ea.) pkgs. crescent rolls
4 oz. mozzarella cheese

Cook ground beef, onion, and green pepper until done; drain. Stir in garlic powder and spaghetti sauce. Cool. Unroll 1 pkg. crescent rolls. Seal perforations. Roll to an 11x9-inch rectangle. Arrange ½ of cheese in center. Cover with ½ of meat mixture. Starting at long edge, roll up. Pinch seams together and place seam side down on cookie sheet. Repeat with other package of rolls. Bake at 350° for 30 minutes. Let stand 5 minutes. Yields: 6 servings.

## STUFFED GREEN PEPPERS

6 large green peppers
1 stalk celery (chopped)
1 clove garlic (minced)
1 lb. lean ground beef
1 (4 oz.) can sliced mushrooms
3 C. cooked rice

1 tsp. salt
¼ tsp. pepper
Dash of cayenne pepper
1 (10¾ oz.) can condensed cream
  of mushroom soup
2 chicken bouillon cubes

Cut a thin slice from the stem end of each pepper. Remove seeds and ribs. Blanch peppers in boiling water for 2-3 minutes; drain. Saute onions, celery, garlic and meat until all trace of pink is gone. Drain mushrooms, reserving liquid. Stir in mushrooms, rice, salt, pepper, and cayenne. Add enough water to mushroom liquid to measure ¾ C. Blend with soup and bouillon cubes. Heat. Stir ½ C. of soup mixture into rice mixture. Stuff peppers with rice mixture and place in a shallow pan. Pour ½ C. water around peppers. Bake at 350° for 20-25 minutes or until heated through. Serve with remaining soup mixture, heated, as a sauce. Serves 6.

# Bean Salad

2 C. cooked dry beans (drained)
1 C. sliced sweet pickles
1 diced onion
Salt and pepper
1 C. vinegar
1 C. sugar
1 egg
1 tsp. cornstarch

Cool dry beans. Mix vinegar, sugar, egg, and cornstarch; cook until thick. Cool. Pour over beans, pickles, and onion. Mix together and add salt and pepper, to taste.

# Green Beans and Sausage

1 lb. green beans
1 (12 oz.) pkg. link sausages
2 T. chopped onion

Clean and snap beans. Place in kettle and cover with water. Add onion and cut-up sausages. Cover and simmer until beans are barely tender. Serves 3 to 4.

## CREAMED CUCUMBERS

3 medium cucumbers (sliced)
1 tsp. salt
4 T. vinegar
8 oz. sour cream

Dash salt
1 medium onion (sliced)
½ C. sugar
½ tsp. dill weed
Dash pepper

Combine cucumbers, onion, and salt. Cover with water and let stand at least 2 hours. Mix vinegar, sugar, sour cream, dill weed, salt and pepper. Drain cucumbers and add mixture.

## Butter Beans

2⅔ C. dried butter beans or lima beans
3 qts. water

½ C. chopped onion
1 tsp. salt
6 strips bacon or ham hocks

Soak beans in water overnight. Or, bring to a boil, simmer 3 minutes, and let stand for 1 hour. Rinse, cover with water; add onion and bacon (or ham hocks). Simmer, uncovered for about 30 to 40 minutes until beans are tender. Add salt and simmer another 4 to 5 minutes.

## Broccoli and Cauliflower Salad

1 head of cauliflower
1 bunch of green onions
1 bunch of broccoli
1 C. raw frozen peas
2 C. salad dressing

Splash of buttermilk
2 tsp. salt
2 C. sour cream
3 tsp. garlic

Wash and cut into pieces the cauliflower, onions, broccoli, and peas. Mix salad dressing, buttermilk, salt, sour cream, and garlic. Mix well with vegetables.

## Cold Green Beans Sesame

½ lb. green beans
Water
1 T. sugar

2 T. soy sauce
4 T. sesame seeds

Cut green beans on the diagonal. Cook for about 10 minutes in water until beans are not quite tender. Smash sesame seeds with side of cleaver. Place in bowl; add sugar and soy sauce, stirring until sugar is dissolved. Add green beans to bowl and toss well.

## Cauliflower Salad

1 large head cauliflower
1 small chopped onion
1 (10 oz.) pkg. frozen peas
1 pt. salad dressing

1½ T. Schilling salad dressing spice
1 (3 oz.) pkg. shredded cheese
8 slices bacon (fries crisply)

Mix in order given. Set in refrigerator overnight. Eat next day.

## Cauliflower and Pea Salad

3 stalks celery (chopped)
1 small onion (grated)
1 C. mayonnaise
1½ tsp. seasoned salt

¾ tsp. milk or more
20 oz. bag frozen peas (thawed)
1 head of cauliflower (small pieces)

Mix chopped celery, grated onion, thawed peas, and a head of cauliflower (cut in small pieces). Add mayonnaise, salt, and milk; let marinate for several hours.

# Cauliflower with Cheese Sauce

1 cauliflower  
⅓ C. mayonnaise  
2 tsp. prepared mustard  

2 C. Cheddar cheese, shredded  
2 T. chopped parsley  

Cook cauliflower in boiling water until tender; drain well. Spread cauliflower with combined mayonnaise and mustard. Top with cheese. Place in 350° oven for 10 minutes or until cheese melts. Serves 6 to 8.

## Sour Cream Eggplant

1 large or 2 medium eggplants  
French dressing  
1 or 2 cloves garlic  

Sour cream  
Chives or green onions with tops  

Smash garlic and place in jar with French dressing. Peel eggplants and slice into ¾-inch slices. Marinated for 1 hour in French dressing.

Bake at 450° for 25 minutes. Remove eggplant from oven and spread with sour-cream chive mixture. With oven door open, heat an additional 5 minutes.

## Pan Fried Eggplant

1 eggplant  
1 egg  
Flour  

½ tsp. salt  
Bacon fat or oil  

Peel and slice eggplant. Soak for 30 minutes in salty water; drain. Dip slices in egg, then in flour. Fry in hot fat until brown. Turn and brown other side. Place on paper towels. Serves 5 to 6.

## Pumpkin Cookies

1 C. shortening  
2 C. sugar  
2 C. cooked pumpkin  
2 tsp. vanilla  
4 C. flour  

2 tsp. baking soda  
2 tsp. cinnamon  
1 (12 oz.) pkg. chocolate chips  
   and nuts (optional)  

Cream shortening and sugar. Stir in pumpkin and vanilla. Sift flour, soda, and cinnamon. Stir into mixture. Add chocolate chips and nuts. Drop by spoonfuls onto greased cookie sheet. Bake at 375° for 12-15 minutes. Makes 48 cookies.

## BEET PICKLES

Wash and snub beets leaving part of the tops and root on (so they don't bleed). In a large pan or kettle, cover beets with water and cook until tender. Cool, peel and quarter or slice beets. Place beets in a boiling solution of:

1 C. vinegar  
1 T. pickling spice (tied in a bag or loose)  
1 C. water  
1 C. sugar

Boil 5-10 minutes. Place in pint or quart jars, covering with liquid. Seal. Hot bath for 20 minutes. Prepare more solution as needed.

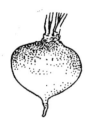

## SPINACH SALAD

1 bunch fresh spinach  
1 medium onion  
Crispy bacon bits  
½ C. vinegar  
½ C. sugar  
¼ C. salad oil  
2 hard boiled eggs (opt.)

Wash spinach, drain. Tear in smaller pieces. Cut up onion in small pieces. Add about 1 C. bacon bits. Combine vinegar, sugar, and oil. Shake well and pour over spinach. Hard boiled eggs are optional. Serve immediately.

## EGGPLANT CREOLE

1 eggplant (peeled, diced, cooked)  
3 T. butter  
3 T. flour  
1 T. salt  

3 large tomatoes or 2 C.  
1 bell pepper (chopped)  
1 onion (chopped)  
1 T. brown sugar  

Mix together all ingredients except eggplant and cooked for 5 minutes. Put eggplant in casserole and pour sauce over the top. Sprinkle with bread crumbs and cheese. Bake at 350° about 30 minutes.

# Mashed Parsnips

2 C. peeled & chopped parsnips  
¼ tsp. salt  

2 T. milk or cream  
2 T. butter, divided  
Dash of nutmeg  

Simmer parsnips in boiling salted water for 15 to 20 minutes until tender. Drain and mash with potato masher or blender. Stir in cream and 1 T. butter. Pour into warm serving dish and top with 1 T. of butter and a brief sprinkle of nutmeg. Serves 2 to 3.

# Turnips Turnips Turnips

3 C. sliced turnips
3 T. butter
1 T. sugar

1 T. parsley, chopped
3 T. water

Melt butter in skillet. Add turnips, water, and sugar. Stir well. Cover and cook, stirring often for about 6 minutes or until turnips are tender. Remove lid and cook until almost dry. Serves 5 to 6.

# Swede Potatoes

1 medium rutabaga
3 T. butter
Water

¼ tsp. salt
Optional: 2 tsp. chopped parsley

Peel rutabaga and cut into pieces about ½-inch square. Put in saucepan with salt and water to cover. Cook, covered for about 20 minutes until tender. Add butter and sprinkle with parsley, if desired.

## Knobby Kohlrabi

1 T. peanut or other oil
1 lb. kohlrabi, peeled & diced
½ tsp. sugar
2 T. chopped fresh parsley
½ C. chicken stock or bouillon

Place oil in skillet over low heat. When hot, add all ingredients and cook, covered for 25 to 30 minutes until kohlrabi is tender.

## French Fried Onion Rings

1 large onion
1 C. cornmeal
2 eggs, beaten
⅔ C. milk
Oil for frying

Cut cleaned onion into ½-inch slices; separate rings. Soak in cold water for 20 minutes. Beat together cornmeal, eggs, and milk. Drain onions, dredge with cornmeal, dip into batter. Fry in hot oil and drain on brown paper or paper towels. Serve hot.

## Oriental Casserole

1½ lbs. lean hamburger
2 C. chopped onion
1 C. diced celery

½ tsp. pepper
⅓ C. raw rice

1 small can water chestnuts
(cut up)
1 small cut mushrooms

⅓ C. soy sauce

Drain liquid off chestnuts and mushrooms and save. Brown hamburger with onion and celery. Add the pepper, soy sauce, and rice. Take drained juice and add enough water to make 2 C. liquid. Bring to a boil and pour over the hamburger mixture to which the mushrooms and chestnuts have been added. Pour in buttered 2-qt. casserole. Bake 1 hour at 350° in covered dish.

## Snow Pea Soup

3 C. chicken stock with pieces
of chicken
4 T. chopped green onions

¼ C. chopped carrot
1½ C. snow peas, cut into
bite-size pieces

Combine all ingredients. Bring to a boil and simmer over low heat until snow peas are tender, 15 to 20 minutes. Salt, to taste. Serves 4.

## Thousand Island Dip for Raw Vegetables

8 oz. pkg. cream cheese
⅓ C. Thousand Island salad dressing
2 T. catsup
3 tsp. minced onion

Combine all of the ingredients and chill. Serve with raw sliced carrots, celery, green peppers, cucumbers, and mushrooms.

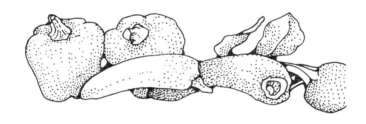

## Red Radish Relish

35 small red radishes
1½ tsp. seasoned salt
½ C. sour cream
¾ tsp. vinegar
2 T. chives

Chop radishes or put in blender with a little water. Buzz on medium unt¹ radishes are well chopped. Place paper towels in colander and pour in radishes; allow to drain for 1 hour. Press to extract as much water as possible.

In bowl combine cream, vinegar, and chives. Stir to blend. Add radishes and refrigerate. Serve with meat or pile on snack crackers.

## MIXED MUSTARD PICKLES

10 medium cucumbers (sliced)
2 C. sliced onion
3 green peppers (sliced)
1 qt. green tomatoes (sliced)
1 small head cauliflower (cut in clowerets)

½ C. salt
4 C. sugar
⅔ C. dry mustard
½ C. flour
1 T. turmeric
2 C. cider vinegar
2 C. water

Place vegetables in a large bowl, sprinkle with salt and let stand overnight. Drain well. Make sauce by mixing sugar, mustard, flour, and turmeric, then stirring in vinegar and water and cooking until smooth and thickened. Add vegetables to boiling sauce and cook until just tender, stirring occasionally to prevent sticking. Fill hot sterilized jars with pickles and sauce and seal at once.

## CHEESY VEGETABLE SOUP

3 T. butter or margarine
3 T. all-purpose flour
2 (14½ oz. ea.) cans chicken broth
2 C. coarsely chopped broccoli
¾ C. chopped carrots
½ C. chopped celery

1 small onion (chopped)
½ tsp. salt
¼ tsp. garlic powder
¼ tsp. dried thyme
1 egg yolk
1 C. heavy cream
1½ C. (6 oz.) shredded Swiss cheese

Melt butter in a heavy 4-qt. saucepan; add flour. Cook and stir until thick and bubbly; remove from the heat. Gradually blend in broth. Add next 6 ingredients; return to the heat and bring to a boil. Reduce heat; cover and simmer for 20 minutes or until vegetables are tender. In a small bowl, blend egg yolk and cream. Gradually blend in several tablespoonfuls of hot soup; return all to saucepan, stirring until slighlty thickened. Simmer for another 15-20 minutes. Stir in cheese and heat over medium until melted. Yield: 8-10 servings (2½ qts.)

## TACO SALAD

1 lb. hamburger
1 small chopped onion
1 head chopped lettuce
2 tomatoes (chopped)

1 C. shredded cheddar cheese
1 C. Miracle Whip
2 T. dry taco dressing
2 C. crushed Doritos

Mix Miracle Whip and taco seasoning and chill. Brown hamburger and onion; drain. Cool. Mix all ingredients together and pour chilled dressing on top to serve.

## SUPER SALAD

1 lb. macaroni (cooked)
1 large green pepper (chopped)
4 shredded carrots
1 small onion (chopped)
1 can Eagle Brand condensed milk

1 C. sugar
1 C. vinegar
2 C. mayonnaise
1 tsp. salt
¼ tsp. pepper

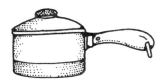

Mix the last six ingredients, than add the veggies and macaroni.

## MIXED VEGETABLE CASSEROLE

1 lb. hamburger
½ C. chopped onions
2 C. diced potatoes (raw)
1½ C. diced carrots (raw)
16 oz. can string beans
1 C. diced celery
1 (10¾ oz.) can cream of celery soup

Brown hamburger and onion; drain. Mix rest of ingredients in with meat. Pour into greased 2-qt. casserole. Bake at 350° for 1½ hours or until vegetables are tender.

## WITCH'S BREW

1½ lbs. hamburger (browned and drained)
1 green pepper (diced)
2 large onions (chopped)
5 sticks celery (diced)
1 (28 oz.) can whole tomatoes
½ tsp. salt
1 (15 oz.) can kidney beans
1 (4 oz.) can mushrooms (stems and pieces)
¼ lb. bacon (fried and cut into pieces)
4 C. cooked noodles (drained)
¼ tsp. pepper

Mix altogether and bake at 350° for 40 minutes.

## ONE DISH MEAL

2 large potatoes (diced)
2 carrots (sliced)
1 small diced onion
½ C. uncooked rice
1½ C. tomato juice

1 lb. ground beef (cooked and drained)
1 tsp. salt
½ tsp. pepper

Layer in greased casserole in order given. Bake at 350° for 1½ hours.


## OVEN STEW

2 lbs. beef stew meat
1 bunch carrots (chunked)
4 potatoes (chopped)
1 C. celery (sliced)
¼ C. green pepper (diced)
2 large onions (chopped)

1 T. salt
Pepper to taste
1 T. sugar
3 T. minute tapioca
2 C. well drained tomatoes

Layer as written. Bake at 250° for 5-6 hours. Uncover last hour.

## STEAK IN A CROCK POT

3 lbs. round steak
1 (10¾ oz.) can beef vegetable soup
1 (10¾ oz.) can cream of mushroom soup)

1 (10¾ oz.) can tomato soup
½ large onion (chopped)
4 carrots (sliced)
2-3 potatoes (sliced)

Trim steak and cut in bite-size pieces. Mix 3 cans of soup together. In greased 3-qt. crock pot, put ¾ C. soup mixture in bottom. Dip pieces of meat in remaining soup mixture and lay in crock pot, pouring any leftover soup on top of meat. Add sliced carrots; then sliced potatoes on top. Cook on crock pot on high for 5 hours.

## VEGETABLE DIP

3 oz. cream cheese (soft)
2 tsp. milk
2 T. ketchup

2 tsp. French or Western dressing
1 tsp. minced onion

Mix together and chill to blend flavor. Serve with your favorite vegetables.

## COLE SLAW

1/8 C. vinegar
1/8 C. oil
½ C. sugar

½ C. Miracle Whip
1 tsp. salt
Shredded cabbage
Shredded carrots

Beat all ingredients until well blended. Mix with cabbage. Stir in carrots.

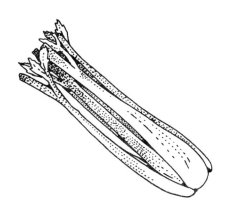

## RAW VEGETABLE SALAD

1 C. green onions (chopped)
1 C. sliced radishes
1 C. cauliflower (broken)
2 C. shredded carrots
2 C. celery (diced)
2 C. peas (drained)

1 C. sugar
⅓ C. oil
⅔ C. vinegar
1½ tsp. salt
½ tsp. pepper

In large bowl, mix all vegetables. In small bowl mix last 5 ingredients. Pour over bowl of vegetables and mix well. Let stand overnight in refrigerator.

## CREAMED VEGETABLES

Stretch your vegetables with a sauce. Use one vegetable or try a mixture of two or three. Remember this trick when you have to set that extra place on short notice.

1 T. butter or margarine
1 T. flour
¼ tsp. salt

1/8 tsp. pepper
1 C. milk
2 C. cooked vegetable(s)

Melt butter in medium saucepan over low heat. Blend in flour and seasonings. Cook over low heat, stirring until mixture is smooth and bubbly. Remove from heat. Stir in milk. Heat to boiling, stirring constantly. Boil and stir 1 minute. Stir vegetable(s) into sauce; heat through. Makes 4-6 servings.

VARIATIONS:
Au Gratin Vegetables: Heat oven to 325°. Add 1 tsp. dry mustard to flour mixture and stir 1 C. shredded process American cheese into hot cream sauce. Cook, stirring constantly, until cheese is melted. Pour vegetable mixture into ungreased 1-qt. casserole; sprinkle with ½ C. cereal or crumb topping. Bake uncovered 15 minutes or until heated through.

Scalloped Vegetables: Heat oven to 325° and pour vegetable mixture into ungreased 1-qt. casserole; sprinkle with ½ C. cereal or crumb topping. Bake uncovered 15 minutes or until heated through.

Vegetables in Cheese Sauce: Add 1 tsp. dry mustard to flour mixture and stir 1 C. shredded process American cheese (about 4 oz.) into hot cream sauce. Cook, stirring constantly, until cheese is melted.

## BROCCOLI AND CAULIFLOWER SALAD

1 bunch broccoli  
1 head cauliflower  
2 bunches green onions  
6 slices bacon (fried and crumbled)

DRESSING:  
½ C. mayonnaise  
¼ C. sugar  
⅓ C. wine vinegar  
⅓ C. oil

Mix dressing with vegetables and bacon and let stand in refrigerator for 24 hours. Toss before serving.

## CAULIFLOWER AND PEA SALAD

1 C. Hellmann's mayonnaise  
1 pkg. Ranch original salad dressing mix  
1 C. chopped celery  
2 C. frozen peas  
1½ C. raw cauliflower pieces

Mix mayonnaise, dressing mix and celery. Pour over peas and cauliflower. Mix well. Chill and serve. (Tastes better when made the day before.)

## Rice Casserole

½ C. raw rice
1 can cream of chicken soup
½ C. water (more may have to be added as it cooks)
3-4 C. mixed vegetables (chopped) (celery, broccoli, carrots, onion-small amount, green pepper)

Combine above ingredients. Bake at 350° in a greased casserole for about 1 hour and 15 minutes. It is done when rice is soft.

## QUICK VEGETABLES AU GRATIN
*(Literally, "au gratin" is French for "with crumbs". It usually means the addition of cheese to scalloped vegetables.)*

Heat oven to 325°. Stir together 1 (10¾ oz.) can condensed cheddar cheese soup, ½ tsp. Worcestershire sauce, and ½ tsp. dry mustard. Gently stir into 2 C. cooked vegetable(s). Pour mixture into ungreased 1-qt. casserole. Sprinkle with ½ C. cereal or crumb topping. Bake uncovered 15-20 minutes or until heated through. Makes 4-6 servings.

## Broccoli-Cauliflower Soup

1 lb. fresh broccoli (cut in small pieces)
1 lb. fresh cauliflower (cut in small pieces)
2 chicken bouillon cubes
½ C. margarine
1 medium onion (chopped)
½ C. flour
4 C. milk
2 C. shredded cheddar cheese

Boil broccoli, cauliflower, and chicken bouillon cubes in a 3-qt. saucepan with just enough water to cover, until vegetables are just tender, approximately 3 minutes. Meanwhile, saute ½ C. margarine and 1 medium chopped onion in a ½-qt. saucepan. Add ½ C. flour and mix. Add 4 C. milk and cook over medium heat until thickened. Add mixture to vegetables, after they have been boiled. Add 2 C. shredded cheddar cheese. Heat until cheese melts through. Season with salt and pepper and serve. Yield: Serves 4-6.

## Cheese Soup

1 C. water
1 large potato (shredded)
1 medium onion (chopped)
1 medium carrot (grated)
1 stalk celery (finely chopped)
1 C. chicken consomme or broth (or use 2 tsp. chicken bouillon dissolved in 1 C. hot water)
½ C. Half and Half
1½ C. shredded sharp cheddar cheese (about 6 oz.)

Combine water, potato, onion, carrot, and celery in 2-qt. casserole; cover. Microwave at high (100%) until potatoes are tender, 12-17 minutes, stirring after half the cooking time. Stir in consomme and Half and Half; cover. Microwave at medium-high (70%) until heated through, 6-8 minutes. Mix in cheese, stirring until melted.

## Cauliflower and Broccoli Salad

1 head cauliflower (broken in small pieces)
1 bunch broccoli (broken in pieces)
1 reg. pkg. Hidden Valley seasonings
8 radishes (sliced)
1 C. salad dressing
1 C. sour cream

Place vegetables in a large bowl. Combine salad dressing, sour cream, and package Hidden Valley seasonings and pour over vegetables. Mix well.

## Broccoli & Cauliflower Salad

1 head cauliflower
1 pkg. radishes
2 C. frozen peas
2 T. dried onion flakes
1 C. mayonnaise
¼ C. milk
1 pkg. Hidden Valley original dressing

Break cauliflower and add peas, radishes, and onion flakes; set aside. Mix mayonnaise, milk, and dressing. Combine with vegetables. Pour into pan and refrigerate.

## Vegetable Combo

1 head of cauliflower (chopped)
1 bunch of broccoli (chopped)
2 carrots (sliced)
15 radishes (sliced)
2 cucumbers (sliced)

1 C. sour cream
1 C. mayonnaise
1 pkg. garlic salad dressing mix or Hidden Valley Dressing
(mixed according to pkg.)

Combine cauliflower, broccoli, carrots, radishes, and cucumbers; chill. Mix sour cream, mayonnaise, and garlic salad dressing. Add this dressing just before serving.

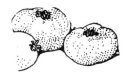

## Refrigerator Slaw

1 medium head cabbage (shredded)
1 tsp. salt
2 carrots (grated)

2 stalks celery (chopped)
½ green pepper (chopped)

DRESSING:
1 C. vinegar
½ C. water
1½ C. sugar

1 tsp. celery seed
1 tsp. mustard seed

Mix shredded cabbage and salt. Let stand for 1 hour. Drain liquid. Add rest of vegetables. Boil dressing ingredients for 1 minute. Cool. Pour over vegetables mixture. Refrigerate. Will keep up to 1 week.

## Garden Salad

2 C. diced cauliflower  
2 C. diced celery  
2 C. shredded carrots  
1 C. sliced radishes  
1 C. drained peas  

1 C. vinegar  
¼ C. salad oil  
½ C. sugar  
1 tsp. salt  
¼ tsp. pepper  

Mix all vegetables toether. Mix vinegar, oil, sugar, salt, and pepper. Fold into vegetables and refrigerate. Can be made 2 days ahead of time for use.

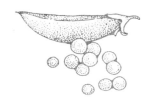

## Broccoli and Cauliflower Salad

2 C. chopped broccoli  
2 C. chopped cauliflower  
1 C. chopped pepper (half red)  
1 C. chopped celery  
1 C. chopped & seeded tomatoes  
1 C. cubed cheese (cheddar)  

6 green onions (chopped)  
1 C. mayonnaise  
1 C. sour cream  
1 T. lemon juice  
1 T. powdered sugar  

Combine broccoli, cauliflower, peppers, celery, tomatoes, cheese, and onions. Blend together mayonnaise, sour cream, lemon juice, and powdered sugar. Pour over vegetables and toss until well coated. This makes a large salad and will serve 15 to 20 people.

## Dill Pickles

1 C. sugar  
1 C. vinegar  
2 T. salt  

Onion  
Dill  

Boil first 3 ingredients. Cut onion and put a slice in the bottom of the jar. Also head of dill. Fill jar with sliced cucumbers. Cover with above mixture and seal. Make sure syrup is real hot.

## Dill Pickles

Wash cucumbers and pack in hot jars. Add 1 or 2 heads of dill to each jar. Combine 2 qts. water, 1 qt. vinegar, and 1 C. pickling salt and boil hard. Pour over cucumbers and seal jars.

## PUMPKIN BARS

1 C. oil
4 eggs
2 C. flour
2 tsp. baking powder
½ tsp. salt

2 C. sugar
2 C. cooked pumpkin
2 tsp. cinnamon
1 tsp. baking soda

Mix together oil and sugar. Add eggs and pumpkin. Sift together flour, cinnamon, baking powder, baking soda, and salt. Add flour mixture to pumpkin mixture. Place in greased jelly roll or cake pan and bake at 350° for 25-30 minutes.

FROSTING:
1 (3 oz.) pkg. cream cheese (softened)
¾ stick oleo

1 tsp. vanilla
1¾ C. powdered sugar
1 T. milk or cream

Mix together cream cheese, sugar, oleo, milk, and vanilla in order. Spread on cooled bars.

# Frosted Pumpkin Bars

2 C. flour
2 C. sugar
1 tsp. baking powder
1 tsp. soda
½ tsp. salt
2 tsp. cinnamon
4 eggs
1 C. salad oil

1 C. pumpkin
1 C. nuts
3 ozs. cream cheese
6 T. oleo
1 tsp. milk
1 tsp. vanilla
¾ lb. powdered sugar

Combine flour, sugar, baking powder, soda, salt, and cinnamon. Make a well in the center. Add the eggs, oil, pumpkin, and nuts; mix until well mixed. Pour into 10x15-inch pan which is greased and floured. Bake at 350° for 20 to 25 minutes.

For Frosting: Mix cream cheese, oleo, milk, vanilla, and powdered sugar until creamy. Frost the cooled bars.

# Pumpkin Bars

**MIX WELL:**
- 2 C. pumpkin
- 2 C. sugar
- 1 C. cooking oil
- 4 eggs

**SIFT TOGETHER:**
- 2 C. flour
- 2 tsp. soda
- 1 tsp. baking powder
- 1 tsp. salt
- 2 tsp. cinnamon

Add flour mixture to pumpkin mixture. Fold in 1 C. each of raisins and nuts, if desired. Bake in 9x13x2-inch greased pan at 350° for 30 to 35 minutes.

**SUGGESTED ICING:**
- 3 ozs. cream cheese
- 6 T. margarine
- 1 tsp. vanilla
- 1 tsp. milk
- 2 C. powdered sugar

Mix well and spread on cooled bars.

# Frozen Pumpkin Pie

- 1 C. pumpkin
- 1 box vanilla instant pudding
- ½ C. milk
- 1 tsp. pumpkin pie spice
- 2½ C. Cool Whip
- 1 graham cracker crust

Using wire whisk; blend pumpkin, pudding, milk, and spice together for 1 minute. Fold in Cool Whip and freeze.

## Pumpkin Roll

¾ C. flour
1 tsp. baking powder
2 tsp. cinnamon
1 T. pumpkin pie spice

3 eggs
1 C. sugar
⅔ C. pumpkin
1 C. nuts

FILLING:
1 C. powdered sugar
1 (8 oz.) pkg. cream cheese
4 T. oleo

1 tsp. vanilla
1 C. whipped topping

Beat eggs and 1 C. sugar until thick; stir in pumpkin. Add dry ingredients. Grease a 15x10x1-inch jelly roll pan. Line with wax paper; grease and flour. Pour in cake mixture and sprinkle with chopped nuts. Bake at 350° for 15 minutes. Invert on towel dusted with powdered sugar; peel off paper. Roll-up towel and cake together. Cool completely. Make cream cheese filling, unroll cake, and cover with filling. Reroll and refrigerate until ready to serve. Slice to serve.

# Cinnamon Pickles

2 gallons cucumbers
2 C. lime
8½ qts. water

Let stand 24 hours. Drain and rinse. Soak 3 hours in cold water. Drain well.

*IN LARGE KETTLE:*
1 bottle red food color
1 T. alum
1 C. vinegar
Water to cover cucumbers

Add cucumbers and simmer 2 hours. Drain.

*SYRUP:*
2 C. vinegar
2 C. water
10 C. sugar
1 pkg. cinnamon candies or red hots

Pour hot over cucumbers. Let set overnight. Repeat for 3 days. On third day pack in jars. Pour hot syrup over and seal.

Citrus Dressing ........................ 160
Easy Honey Cookies .................... 157
Honey Chicken ....................... 159
Honey Milk Balls ..................... 158
No Cook Honey Butter Balls ............ 158
Orange Honey Glazed Ham ............. 160
Pure Honey Basic Cake ................. 157
Sioux Bee Refrigerator Cookies .......... 159

## PURE HONEY BASIC CAKE

1 C. butter
1½ C. honey
2 eggs
1 C. milk

3 C. flour
2 tsp. baking powder
½ tsp. soda
1 tsp. vanilla

Cream together butter and honey. Add egg and beat until smooth. Add milk with dry ingredients. Bake at 350° in 9x13-inch cake pan for 35-40 minutes.

To make a fruit-filled cake, add cherry or apple pie filling over the batter before baking.

Pineapple Upside-Down Cake: On bottom of pan, spread ½ C. butter melted and 1 C. honey. Add 1 large can pineapple. Pour basic cake batter on top. Bake at 350° for 35-40 minutes. When baked, turn cake over on platter.

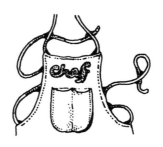

## EASY HONEY COOKIES

1 C. shortening
3 eggs
2½ tsp. soda
7/8 C. milk
2½ C. oatmeal

1¼ C. honey
1 tsp. cinnamon
2 tsp. vanilla
3 C. flour
1 pkg. chocolate chips

Cream together shortening, honey, and eggs. Add and mix all remaining ingredients. Let stand for 10 minutes. Bake at 325° for 8-10 minutes. Makes 5-6 dozen.

## HONEY MILK BALLS

½ C. honey
½ C. peanut butter

1 C. nonfat dry milk
1 C. uncooked quick oats

CHOCOLATE COATING:
6 blocks bitter chocolate
1 T. butter

½ block wax
1 tsp. vanilla

Combine honey and peanut butter. Work in dry milk and oats. Refrigerate until cool, then roll into small balls. Dip into chocolate coating.

## No Cook Honey Butter Balls

1 C. peanut butter
1 C. honey
2 C. powdered milk, dry

¾ C. nuts (chopped)
1 C. corn flakes, crushed

Stir together peanut butter and honey. Stir in dry milk and nuts. Shape into balls. Roll in corn flake crumbs. Store these in the refrigerator.

## HONEY CHICKEN

4 whole chicken breasts (boned, skinned, and split)
3 T. butter or margarine
1 small onion
¼ C. orange juice

1 C. chicken broth
½ C. honey
¼ tsp. pepper
2 tsp. soy sauce
1 (8 oz.) can pineapple chunks

In large skillet brown chicken breasts in butter. Remove and set aside. Brown chopped onion. Add next 5 ingredients and bring to a boil. Reduce heat and add chicken. Cover and simmer 40 minutes. Add pineapple and continue to simmer for a few minutes before serving.

## Sioux Bee Refrigerator Cookies

½ C. peanut butter
¼ C. honey
1 C. flaked coconut

2 C. puffed wheat, sugar-coated, divided ½ C. & 1½ C.
⅓ C. chopped nuts

Combine peanut butter, honey, and coconut; stir well. Add ½ C. puffed wheat and stir in. Spread remaining cereal on waxed-paper lined plate. Shape mixture into balls with hands and roll in cereal. Refrigerate until firm.

## Citrus Dressing

4 T. fresh lemon juice
2 T. fresh orange juice
1 T. honey

1 T. olive oil
1½ tsp. grated orange peel

Combine all ingredients and refrigerate. Use on fruit salads or greens. Shake before using.

## Orange Honey Glazed Ham

1 (3 lb.) canned ham
1 (3 oz.) pkg. orange gelatin

2 T. onion, chopped
1 tsp. ground cloves
1 T. honey

Drain ham; rub surface with cloves. Place ham on a large sheet of heavy-duty aluminum foil. Place on rack in baking pan. Sprinkle dry gelatin over ham and drizzle with honey. Close foil with "grocery store" fold. Bake at 350° for 1 hour. Open foil, being careful to avoid steam, and bake another 30 minutes. Serves 8 to 10.

# NUTS

Almond Toffee .......................... 176
Black Walnut Clusters .................. 169
Brown Sugar Pecans ..................... 168
Butter Pecan Squares ................... 179
Butterscotch Dessert ................... 173
Cherry Nut Fudge ....................... 168
Chocolate Cherry Creams ................ 185
Chocolate Pecan Bars ................... 178
Christmas Surprise Dessert ............. 175
Date Nut Cookies ....................... 165
Deluxe Pecan Pie ....................... 181
Florentine Sundae ...................... 171
Garlic Pecans .......................... 170
Ginger Pops ............................ 166
Hazelnut Butter Cookies ................ 167
Ice Box Butterscotch Cookies ........... 183
Layered Praline Pie .................... 181
Lisa's Rum Balls ....................... 165
Microwave Chocolate Nut Squares ........ 166
Microwave Chocolate-Pecan Fudge ........ 177
M-M-M-M Dessert ........................ 172
Never Tell Salad ....................... 170
Nut Cookies ............................ 182
Peanut Clusters ........................ 175
Pecan Pie .............................. 163
Pecan Pie .............................. 164
Pecan Pie .............................. 184
Pecan Pie .............................. 184
Pecan Pie Bars ......................... 179
Pecan Rolls ............................ 180
Pecan Tarts ............................ 173
Pecan Tarts ............................ 183
Pecan Waffles .......................... 169
Philly Pecan Logs ...................... 176
Pistachio Cream Pie .................... 163
Pudding Cookie Surprise ................ 174
Pumpkin Delight ........................ 172
Pumpkin Pecan Bread .................... 185
Raspberry Salad ........................ 171
Refrigerator Oat Bars .................. 182
Rocky Road Nut Bars .................... 178
Spice Pecans ........................... 164
Super Good Nuts ........................ 167
Walnut Glory Cake ...................... 180

## Pecan Pie

½ C. sugar
3 T. butter
3 eggs (beaten)
1 T. flour

1 C. white Karo syrup
¼ tsp. salt
1 C. pecans
1 tsp. vanilla

Cream butter and sugar; add beaten eggs and mix thoroughly. Add remaining ingredients. Pour into unbaked pie shell. Bake at 350° for about 50 minutes.

### PISTACHIO CREAM PIE

½ C. margarine
2 T. sugar
1½ C. flour

½ C. chopped nuts
1 C. coconut

FILLING:
1 C. powdered sugar
8 oz. cream cheese
16 oz. Cool Whip

2 pkgs. pistachio instant pudding
3 C. milk

Cut margarine into flour and sugar as for crust. Add nuts and pat into bottom of 9x13-inch pan. Bake 15 minutes at 350°. Toast coconut in a separate pan.

For Filling: Beat cream cheese until soft. Add sugar and 1 C. Cool Whip. Spread over cooled crust. Mix milk with pudding. When thickened spread on cheese mixture. Top with remaining Cool Whip. Sprinkle with coconut. Cut into squares. May also be made with chocolate, coconut, or butter pecan pudding.

## Spice Pecans

1 lb. pecan halves  
1 egg white  
1 tsp. water  

½ C. sugar  
½ tsp. cinnamon  
¼ tsp. salt  

Beat egg white and water until frothy. Pour over pecans and stir until pecans coated. In bowl mix sugar, cinnamon, and salt. Put pecans in sugar mixture and stir gently. Bake at 225° in 9x13-inch pan for 1 hour, turning every 15 minutes. Cool and store in container.

## Pecan Pie

3 eggs (slightly beaten)  
1 C. white syrup  
1 C. sugar  
Pinch of salt  

1½ C. pecan meats  
3 T. melted butter  
1 tsp. vanilla  

Mix in order given and bake in raw pie crust about 400° for 45 minutes.

## Lisa's Rum Balls

1 C. heavy cream
1 lb. semi-sweet chocolate, grated
1½ C. chopped walnuts
⅔ C. rum

Bring cream to a boil over medium heat and add grated chocolate; cook and stir with wooden spoon until chocolate is thick and smooth. Remove from heat. When cold, stir in nuts and rum. When firm enough, roll into balls. Roll in grated chocolate or ground nuts, if desired.

## Date Nut Cookies

2 C. chopped pecans
2 C. chopped pitted dates
2 C. coconut (reserve ½ C.)
¾ C. dark brown sugar (packed)
2 eggs (beaten)

Work together nuts, dates, and coconut or put through a food chopper. Combine all ingredients and shape into balls. Roll in reserved coconut. Space 2-inches apart on a greased cookie sheet. Bake at 350° for 10 minutes or until golden brown. Make 3 dozen cookies.

## Ginger Pops

1 pkg. gingerbread mix
⅓ C. warm water

1 C. chopped nuts

Combine all ingredients. Shape dough into 1-inch balls. Place 2-inches apart on ungreased cookie sheet. Grease bottom of large glass and flatten cookies. Place in 375° oven for 8 to 10 minutes. Cool cookies 1 minute before removing with a spatula. Cool on wire rack.

## Microwave Chocolate Nut Squares

2½ C. chocolate chips
1 square (1 oz.) baking chocolate, chopped

1 (14 oz.) can sweetened condensed milk
1 C. chopped nuts

Place chocolates in paper dish and cover with plastic wrap. Microwave 3 to 4 minutes on low; stir. If bits are not melted, microwave briefly and test stir again. When melted, pour into mixer bowl, add condensed milk and nuts. Beat until blended. Pour into plastic-lined 8x8x2-inch pan. Chill until firm. Cut into 1-inch squares.

Store in the refrigerator or freeze. Makes 64 pieces.

## Super Good Nuts

¾ C. brown sugar  
1 unbeaten egg white

2½ C. coarsely chopped nuts

Mix the sugar and unbeaten egg white; stir in the nuts. Smooth on ungreased baking sheet. Bake at 275° for 30 minutes. Remove with spatual and break into pieces.

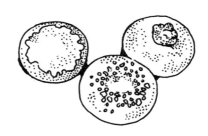

## Hazlenut Butter Cookies

2 C. flour  
1 C. butter (room temp.)  
½ C. powdered sugar

¾ tsp. vanilla  
Chopped hazelnuts  
   (or other nuts)

Combine sugar, butter, flour, and vanilla. Mix well and form into 2-inch long logs. Sprinkle with chopped nuts. Bake at 300° for 10 minutes.

# Brown Sugar Pecans

1 C. brown sugar
½ C. white sugar
½ C. sour cream
1½ tsp. vanilla
2½ C. chopped pecans

Stir together sugars and sour cream in a saucepan. Cook over low heat, stirring constantly to 277° on a candy thermometer, or until a few drops form a soft ball in a cup of cold water. Pour into bowl, add flavoring and nuts. Stir until nuts are well coated. Drop from teaspoon onto aluminum foil or waxed paper.

# Cherry Nut Fudge

2 C. sugar
1 C. heavy cream
¾ C. chopped nuts
½ C. chopped maraschino cherries
1 tsp. vanilla

Combine sugar and cream in saucepan. Bring to boil over medium heat and add salt. Cook, stirring constantly with a wooden spoon, to softball stage, 230° to 240°. Remove from heat. When almost cold, beat until thick and creamy. Cover with a damp cloth for 30 minutes. With hands, work in vanilla, nuts, and cherries. Line pan with waxed paper and press fondant into pan.

## Black Walnut Clusters

1 pkg. butterscotch pudding, dry
1 C. sugar
2 tsp. butter

½ C. evaporated milk
1½ C. chopped black walnuts or other nuts

Combine all ingredients, except nuts. Bring to an easy boil and cook to the soft ball stage which will be 235° to 240° on your candy thermometer. Let cool a big, then beat until it begins to thicken. Stir in nuts and drop from teaspoon onto waxed paper.

## Pecan Waffles

2 C. biscuit mix
1⅓ C. milk
1 egg

2 T. oil or melted butter
½ C. chopped pecans

Combine all ingredients, except nuts. Whip with hand beater or whisk until smooth. Stir in pecans. Pour onto center of hot waffle iron. Cook until steaming stops. Loosen edges with fork and remove carefully. Makes 3 (9-inch) waffles.

# Garlic Pecans

4 C. pecans
½ stick margarine or butter
2 tsp. Worcestershire sauce

¼ tsp. garlic salt
Dash red pepper

**Melt** butter or shortening. Combine all ingredients in a shallow pan. Toast in 250° oven 50 to 60 minutes.

## NEVER TELL SALAD

1 (No. 2) can crushed pineapple
1½ C. miniature marshmallows
3 oz. orange or lime Jello

1 C. cottage cheese
1 C. whipped cream
½ C. chopped walnuts

Drain juice from pineapple and add enough water to make 1 C. Combine juice/water, marshmallows and Jello. Heat until marshmallows are dissolved. Set aside to cool. When nearly set, fold in whipped cream, cottage cheese, pineapple and nuts. Pour into 8-inch square pan and refrigerate until set.

## RASPBERRY SALAD

1 (3 oz.) pkg. raspberry Jello
3 oz. softened cream cheese
1 C. boiling water
½ C. chopped nuts
Whipped topping

1 small can crushed pineapple
   (partly drained)
½ C. maraschino cherries
   (chopped)

Mix dry Jello with cream cheese; add water. Mix well; let set until like syrup. Add rest of ingredients; fold in Cool Whip or whipped topping.

## FLORENTINE SUNDAE

1 (8 oz.) can pineapple slices
   (drained)
1 qt. orange or lemon frozen
   sherbet

Grated semi-sweet chocolate
Chopped nuts
Flaked coconut

Place a pineapple slice in the bottom of four dessert dishes. Top with a scoop or two of sherbet. Sprinkle with grated chocolate, nuts, and coconut. Serve immediately. Yield: 4 servings.

## PUMPKIN DELIGHT

1 pkg. yellow cake mix
4 eggs
¾ C. sugar
⅔ C. milk
½ C. chopped nuts
¾ C. margarine
2 (1 lb. ea.) cans pumpkin
½ C. packed brown sugar
½ tsp. cinnamon
Whipped cream

Reserve 1 C. dry cake mix. Combine remaining cake mix, ½ C. melted margarine, and 1 beaten egg; mix well. Press into greased and floured 9x13-inch pan. Combine pumpkin, 3 eggs, brown sugar, ½ C. sugar, milk, and cinnamon. Spread over batter. Combine reserved 1 C. cake mix, nuts, ½ C. sugar, and ¼ C. margarine; mix until crumbly. Sprinkle over pumpkin mixture. Bake at 350° for 50-60 minutes or until cake tests done. Cool. Serve. Cut in squares topped with whipped cream.

## M-M-M-M DESSERT

1½ C. flour
1½ sticks margarine
½ C. ground nuts
2 pkgs. instant pudding (small)
1 (8 oz.) pkg. cream cheese
1 C. Cool Whip
1 C. powdered sugar
3 C. milk
1 large Cool Whip
½ C. chopped nuts

Mix flour, margarine, and nuts and press into a 9x13-inch pan. Bake at 375° for 15 minutes. Cool. Beat cream cheese, Cool Whip, and powdered sugar. Put over the crust. Beat 2 pkgs. instant pudding (lemon, butterscotch, or chocolate) with 3 C. milk for 2 minutes. Put over the cheese mixture and top with Cool Whip and nuts. Place in the refrigerator for 24 hours.

## BUTTERSCOTCH DESSERT

CRUST:
1 C. flour
½ C. chopped pecans
1 stick oleo (softened)

Mix until crumbly. Bake at 350° for 12-15 minutes in 9x13-inch pan. Cool real good.

2nd Part: Take 8 oz. pkg. softened cream cheese; add 1 C. powdered sugar and 1 C. Cool Whip. Mix together and spread on cooled crust.

3rd Part: Cook 2 small pkgs. butterscotch pudding and pie filling with 3 C. milk. Cook as you would pudding. Cool. Then spread on cheese mixture. Then take 1 C. Cool Whip and spread on pudding; top with 1 C. toasted almonds.

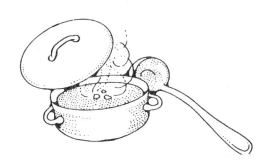

## PECAN TARTS

1 (3 oz.) pkg. cream cheese
1 C. flour
1 stick oleo

FILLING:
1 egg
1 tsp. vanilla
¾ C. brown sugar
⅔ C. chopped pecans

Cream together cream cheese and oleo. Add flour. Mix well. Shape dough* around sides and bottom of greased tart pans. Fill ⅔ full with filling. Bake at 350° for 25 minutes. *About 1½ T.

## PUDDING COOKIE SURPRISE
*(Makes 4 servings)*

1 (4-serving size) pkg. Jello pudding and pie filling (any flavor)
2 C. milk
6 gingersnaps (crumbled)
⅓ C. chopped nuts
2 T. butter or margarine (melted)

Range Top Directions: Combine pudding mix and milk in medium saucepan; blend well. Cook and stir over medium heat until mixture comes to a full boil. Pour into bowl; cover surface of pudding with plastic wrap. Chill. Combine crumbs, nuts, and butter; mix well. Place about 2 T. of the crumb mixture into each of 4 dessert dishes. Stir pudding and spoon into dishes. Sprinkle with remaining crumb mixture.

Microwave Directions:* Combine pudding mix and milk in 1½-qt. microwave-safe bowl; blend well. Cook at High 3 minutes. Stir well and cook 2 minutes longer; then stir again and cook 1 minute or until mixture comes to a boil. Stir; cover surface of pudding with plastic wrap. Chill. Combine crumbs, nuts, and butter; mix well. Place about 2 T. of the crumb mixture into each of 4 dessert dishes. Stir pudding and spoon into dishes. Sprinkle with remaining crumb mixture. *Ovens vary. Cooking time is approximate.

## CHRISTMAS SURPRISE DESSERT

1 stick butter
2 C. oatmeal
½ C. brown sugar

1 C. nuts
1 C. caramel sauce
½ gallon vanilla ice cream

Combine butter, oatmeal and sugar and spread on cookie sheet or bottom of large pan and place in oven at 350° until brown, about 15 minutes. While this is in oven, place jar of caramel sauce in hot water to warm. When baked mixture is cool enough to handle, sprinkle about ¾ of the mixture in bottom of 9x13-inch pan. Sprinkle coarsely chopped nuts over first mixture; drizzle the caramel sauce over the nuts. Place ½ gallon ice cream in biggest mixing bowl and beat. Spread ice cream on top of sauce; sprinkle top with remaining crumb mixture and place dessert in freezer. Freeze several hours before serving. Will serve 21 people. Vanilla wafers or graham crackers may be used in place of oatmeal.

## PEANUT CLUSTERS

12 oz. chocolate chips
12 oz. butterscotch chips

1 lb. English walnuts or peanuts

Melt chips; mix in nuts. Drop by spoonfuls onto waxed paper.

## ALMOND TOFFEE

2½ C. sugar
1 lb. butter
1 tsp. salt

2 tsp. vanilla
6½ T. water
½ lb. raw almonds (ground or cut in small pieces)

Melt butter in large heavy pan. Add other ingredients and boil hard, stirring constantly to 300° (hard crackle). Spread on well buttered cookie sheets. This will make a lot of candy. You will need 3 cookie sheets.) Store in cool dry place.

## PECAN LOGS

2½ C. powdered sugar
1 (3 oz.) pkg. cream cheese
¼ tsp. vanilla
Dash of salt

28 vanilla and chocolate caramels
2 T. water
1 C. chopped pecans

Mix well softened cream cheese and sugar. Stir in vanilla and salt. Shape into four 3-inch logs. Chill overnight. Melt caramels with water in saucepan over low heat. Stir until sauce is smooth. With fork, dip each log in hot caramel sauce and then roll in nuts. Wrap in wax paper and freeze for 15 minutes. Chill in refrigerator until serving time. Slice and serve.

## MICROWAVE FAST CHOCOLATE-PECAN FUDGE

½ C. butter or margarine
¾ C. Hershey's cocoa
4 C. confectioner's sugar

1 tsp. vanilla extract
½ C. evaporated milk
1 C. pecan pieces

Line square pan, 8x8x2-inches, with foil; set aside. In medium micro-proof bowl microwave butter on high (full power) for 1-1½ minutes or until melted. Add cocoa; stir until smooth. Stir in confectioner's sugar and vanilla; blend well (mixture will be dry and crumbly). Stir in evaporated milk. Microwave on high 1 minute; stir. Microwave an additional minute or until mixture is hot. Beat with wooden spoon until smooth; add pecan pieces. Pour into prepared pan. Cover; chill until firm. Cut into squares. Store, covered, in refrigerator. Makes about 4 dozen squares.

For Mint 'N Chocolate Fudge: Follow recipe for Fast Chocolate-Pecan Fudge. Omit pecans. Spread Pastel Mint Topping (below) over chilled fudge.

Pastel Mint Topping: In small bowl cream 3 T. butter, softened; 1 T. water and 1/8-¼ tsp. mint extract. Gradually add 1½ C. confectioner's sugar and 2 drops green food color. Beat until smooth.

## NUT BARS

1½ C. flour  
½ tsp. salt  
1¾ C. sugar  
4 eggs  
1 tsp. vanilla  
1 C. butter (melted)  
1½ C. nuts  
4 C. mini-marshmallows

Beat flour, sugar, salt, eggs, and vanilla. Stir in nuts, then butter. Pour in 13x9-inch greased and floured pan. Bake at 350° for 35-40 minutes. Sprinkle marshmallows over cake while hot, then icing.

ICING:  
1 C. brown sugar  
⅓ C. milk  
¼ C. butter

Boil milk and butter 3 minutes. Add sugar and beat well. Drizzle over marshmallows.

## CHOCOLATE PECAN BARS

1¼ C. flour  
½ C. Hershey's cocoa  
14 oz. sweetened condensed milk  
1 egg  
1 C. powdered sugar  
1 C. (soft) butter  
2 tsp. vanilla  
1½ C. chopped pecans

In a large bowl combine flour, sugar, and cocoa. Cut in butter until crumbly. Mixture will be dry. Press firmly in a 9x13-inch pan. Bake at 350° for 15 minutes. Meanwhile beat the sweetened condensed milk, egg, and vanilla; mixing well. Stir in nuts. Spread evenly over crust and bake 25 minutes longer (or until lightly browned). Cool; cut in squares.

## BUTTER PECAN SQUARES

½ C. butter (softened)
½ C. packed light brown sugar
1 egg
1 tsp. vanilla extract

¾ C. all-purpose flour
2 C. (11.5 oz. pkg.) Hershey's milk chocolate chips (divided)
¾ C. chopped pecans (divided)

Heat oven to 350°. Grease square pan, 8x8x2-inch or 9x9x2-inch. In small mixer bowl cream butter, sugar, egg, and vanilla until light and fluffy. Blend in flour. Stir in 1 C. chocolate chips and ½ C. pecans. Spread into prepared pan. Bake 25-30 minutes or until lightly browned. Remove from oven. Immediately sprinkle with remaining 1 C. chips. Let stand 5-10 minutes or until chips soften; spread evenly. Immediately sprinkle with remaining ¼ C. pecans; press gently onto chocolate. Cool completely. Cut into squares. About 16 squares.

## PECAN PIE BARS

CRUST:
½ C. brown sugar
½ C. butter (melted)

1 C. flour

TOPPING:
2 eggs (slightly beaten)
1 C. brown sugar
1 C. coarsely chopped pecans

1 C. coconut
2 T. flour
½ tsp. baking powder
¼ tsp. salt

Combine crust ingredients and blend well. Press firmly over bottom and sides of an ungreased 7x11x2-inch baking pan. Bake at 375° for 10-15 minutes. Remove from oven. Combine topping ingredients and blend well. Spread over baked crust. Return to oven and bake 15 minutes more. Remove from oven and cool.

# WALNUT GLORY CAKE

¾ C. flour
2 tsp. cinnamon
1 tsp. salt
9 eggs (separated)

1½ C. sugar
2 tsp. vanilla
2 C. finely chopped walnuts

Sift flour with cinnamon and salt. Beat egg whites (1¼ C.) in large mixing bowl until soft mounds form. (Gradually add ¾ C. sugar.) Continue beating until very stiff, straight peaks form (don't underbeat). In a small mixing bowl combine egg yolks, ¾ C. sugar and vanilla. Beat until thick and lemon colored. Stir in dry ingredients. Fold this batter gently but thoroughly into egg whites using a wire whip. Fold in walnuts. Turn into ungreased 10-inch tube pan. Bake at 350° for 55-60 minutes. Cool completely before removing from pan.

# PECAN ROLLS

Raise 1 loaf frozen bread until double in size (about 3 hours). Punch down. Roll out on floured board. Melt 3 T. butter, spread over top; sprinkle white sugar over surface, shake over 1 tsp. cinnamon. Roll in a roll.

MELT:
½ C. margarine
1 T. corn syrup

½ C. brown sugar (firmly packed)
1 T. water

Stir over heat until dissolved. Pour into 8-inch square pan. Put pecans over top of syrup mixture. Cut dough in 12 pieces and put in pan. Cover with dish towel. Wrap in blanket. Let raise about 8 hours. Bake at 350° for 40-45 minutes, or until brown. Makes 12 rolls.

## DELUXE PECAN PIE

3 eggs (slightly beaten)
1 C. Karo (light or dark)
1 C. sugar

2 T. margarine or butter (melted)
1 tsp. vanilla
1½ C. pecans
1 unbaked 9-inch pastry shell

In large bowl stir together first 5 ingredients until well blended. **Stir in pecans.** Pour into pastry shell. Bake at 350° for 50-55 minutes or until knife **comes** out clean. Cool. Serves 8-10.

## LAYERED PRALINE PIE
*(Makes one 9-inch pie)*

⅓ C. butter or margarine
⅓ C. packed brown sugar
½ C. chopped pecans
1 lightly baked 9-inch pie shell*
1¾ C. thawed Cool Whip

1 (6-serving size) pkg. Jello vanilla flavor pudding and pie filling**
2½ C. milk

Combine butter, brown sugar, and nuts in medium saucepan. Heat over medium heat until butter and sugar are melted. Spread in bottom of pie shell. Bake in preheated 450° oven for 5 minutes or until bubbly. Cool on wire rack. Combine pie filling mix and milk in another medium saucepan. Cook and stir over medium heat until mixture comes to a full boil. Pie filling thickens as it cools. Cool 5 minutes, stirring twice. Measure 1 C.; cover with plastic wrap and chill thoroughly. Pour remaining pie filling into pie shell; chill. Blend 1⅓ C. of the whipped topping into measured chilled pie filling. Spoon into pie shell; chill at least 3 hours. Garnish with remaining whipped topping.
*Decrease recommended baking time by 5 minutes. **Substitution: Use 2 (4-serving size ea.) pkgs. Jello vanilla flavor pudding and pie filling, increasing milk to 3½ C.

# Refrigerator Oat Bars

2 C. chocolate chips
¾ C. peanut butter

2½ C. round oat cereal
½ C. chopped nuts

Combine chocolate chips and peanut butter in small saucepan. Stir over very low heat until blended and smooth. Remove from heat; stir in cereal and nuts and pour into waxed paper lined 8x8x2-inch pan. Refrigerate until firm. Cut into 24 squares.

## Nut Cookies

1 C. white sugar
1 C. brown sugar
½ C. soft shortening
2 eggs
1 tsp. vanilla

3 C. flour
1 tsp. soda
½ tsp. salt
1 C. chopped nuts

Heat oven to 375°. Mix well sugars, shortening, eggs, vanilla. Sift together flour, soda, and salt; stir in and add nuts. Shape 1 level tablespoon dough into balls. Place on ungreased cookie sheet. Flatten with bottom of greased glass dipped in sugar. Bake 8-10 minutes. Makes approximately 5 dozen.

# Ice Box Butterscotch Cookies

2 C. brown sugar
1 C. shortening (I use oleo)
2 eggs
1 tsp. cream of tartar
¼ tsp. salt

1 tsp. vanilla
1 tsp. soda
4 C. flour
1 C. nutmeats

Mix ingredients in order given. Make in a roll and let stand overnight. Slice and bake at 350° until nicely browned.

# Pecan Tarts

DOUGH:
1 stick oleo (less 1-inch to be used for filling)
3 oz. cream cheese
1 C. flour

FILLING:
¼ C. sugar
1 inch oleo (1 T.)
1 tsp. vanilla
¾ C. chopped nuts

Mix cream cheese, 1 stick oleo (less 1-inch), and flour. Make 24 balls. Place each ball in a muffin mold (tart-pan size), which has been sprayed with non-stick coating. Mix filling ingredients together. Fill dough and bake at 350° for 30 minutes.

# Pecan Pie

¼ C. margarine
¾ C. sugar
1 tsp. vanilla
2 T. flour
3 eggs

½ C. Kahlua
½ C. dark corn syrup
¾ C. evaporated milk
1 C. pecans
Pie shell

Cream margarine, sugar, vanilla, and flour. Add eggs (1 at a time). Stir in remaining ingredients. Pour into unbaked pie shell. Bake at 400° for 10 minutes. Turn oven down to 325° and bake for 40 minutes more or until set. Makes 1 deep dish pie.

# Pecan Pie

3 eggs
1 C. light corn syrup
½ C. sugar
1 T. margarine (melted)

1 tsp. vanilla
1⅓ C. pecan halves
Pie shell

Beat eggs lightly. Add syrup, sugar, margarine, vanilla, and pecans. Pour into 9-inch unbaked pie shell. Bake at 375° for 40 to 50 minutes.

## Pumpkin Pecan Bread

2 C. sifted flour
½ tsp. soda
1 tsp. cinnamon
1 C. canned pumpkin
½ C. milk
¼ C. softened butter

2 tsp. baking powder
1 tsp. salt
½ tsp. nutmeg
1 C. sugar
2 eggs
1 C. chopped pecans

Sift together dry ingredients. Combine pumpkin, sugar, milk, and eggs in mixing bowl. Add dry ingredients and butter; mix till well blended. Stir in nuts. Spread in well-greased 9×5×4-inch loaf pan. Bake at 350° for 45 to 55 minutes or till toothpick inserted in center comes out clean. Make 1 loaf. May be frozen.

## Chocolate Cherry Creams

1 (6 oz.) pkg. semi-sweet chocolate morsels
½ C. evaporated milk
2½ C. powdered sugar (sifted)

⅓ C. nuts (chopped)
⅓ C. maraschino cherries (cut up)
1¼ C. coconut (cut up)

Put semi-sweet chocolate morsels and evaporated milk into a heavy 2-qt. saucepan. Stir over low heat until chocolate melts completely. Remove from heat. Stir in sifted powdered sugar (mix well), chopped nuts and well drained cut up maraschino cherries. Chill until mixture is firm enough to handle (about 1 hour). Roll teaspoonfuls of mixture into the cut up coconut. Chill until firm, about 4 hours. Makes 30.

# Need a Gift?

## For

- Shower • Birthday • Mother's Day •
- Anniversary • Christmas •

Turn Page For Order Form
*(Order Now While Supply Lasts!)*

## TO ORDER COPIES OF
## The Orchards, Berry Patches and Garden Cookbook

Please send me _____ copies of **Orchards, Berry Patches and Garden** at $11.95 each. (Make checks payable to **Hearts 'N Tummies Cookbooks**.)

Name _____

Street _____

City _____ State _____ Zip Code _____

**SEND ORDERS TO:**
*HEARTS 'N TUMMIES COOKBOOK CO.*
**31798 K18S**
**Sioux City, Iowa 51109**

---

## TO ORDER COPIES OF
## The Orchards, Berry Patches and Garden Cookbook

Please send me _____ copies of **Orchards, Berry Patches and Garden** at $11.95 each. (Make checks payable to **Hearts 'N Tummies Cookbooks**.)

Name _____

Street _____

City _____ State _____ Zip Code _____

**SEND ORDERS TO:**
*HEARTS 'N TUMMIES COOKBOOK CO.*
**31798 K18S**
**Sioux City, Iowa 51109**

Since you have enjoyed this book, perhaps you would be interested in some of these others from QUIXOTE PRESS.

## ARKANSAS BOOKS

HOW TO TALK ARKANSAS
    by Bruce Carlson ............................................. paperback $7.95
ARKANSAS' ROADKILL COOKBOOK
    by Bruce Carlson ............................................. paperback $7.95
REVENGE OF ROADKILL
    by Bruce Carlson ............................................. paperback $7.95
GHOSTS OF THE OZARKS
    by Bruce Carlson ............................................. paperback $9.95
A FIELD GUIDE TO SMALL ARKANSAS FEMALES
    by Bruce Carlson ............................................. paperback $9.95
LET'S US GO DOWN TO THE RIVER 'N...
    by various authors ........................................... paperback $9.95
ARKANSAS' VANISHING OUTHOUSE
    by Bruce Carlson ............................................. paperback $9.95
TALL TALES OF THE MISSISSIPPI RIVER
    by Dan Titus ..................................................... paperback $9.95
LOST & BURIED TREASURE OF THE MISSISSIPPI RIVER
    by Netha Bell & Gary Scholl ......................... paperback $9.95
TALES OF HACKETT'S CREEK
    by Dan Titus ..................................................... paperback $9.95
UNSOLVED MYSTERIES OF THE MISSISSIPPI RIVER
    by Netha Bell ................................................... paperback $9.95
101 WAYS TO USE A DEAD RIVER FLY
    by Bruce Carlson ............................................. paperback $7.95
VACANT LOT, SCHOOL YARD & BACK ALLEY GAMES
    by various authors ........................................... paperback $9.95
HOW TO TALK MIDWESTERN
    by Robert Thomas ........................................... paperback $7.95
ARKANSAS COOKIN'
    by Bruce Carlson .................................. (3x5) paperback $5.95

## DAKOTA BOOKS

HOW TO TALK DAKOTA ............................................. paperback $7.95
Some Pretty Tame, but Kinda Funny Stories About Early
DAKOTA LADIES-OF-THE-EVENING
    by Bruce Carlson ............................................. paperback $9.95

SOUTH DAKOTA ROADKILL COOKBOOK
 by Bruce Carlson ................................................ paperback $7.95
REVENGE OF ROADKILL
 by Bruce Carlson ................................................ paperback $7.95
101 WAYS TO USE A DEAD RIVER FLY
 by Bruce Carlson ................................................ paperback $7.95
LET'S US GO DOWN TO THE RIVER 'N...
 by various authors ............................................. paperback $9.95
LOST & BURIED TREASURE OF THE MISSOURI RIVER
 by Netha Bell .................................................... paperback $9.95
MAKIN' DO IN SOUTH DAKOTA
 by various authors ............................................. paperback $9.95
GUNSHOOTIN', WHISKEY DRINKIN', GIRL CHASIN' STORIES
OUT OF THE OLD DAKOTAS
 by Netha Bell .................................................... paperback $9.95
THE DAKOTAS' VANISHING OUTHOUSE
 by Bruce Carlson ................................................ paperback $9.95
VACANT LOT, SCHOOL YARD & BACK ALLEY GAMES
 by various authors ............................................. paperback $9.95
HOW TO TALK MIDWESTERN
 by Robert Thomas ............................................. paperback $7.95
DAKOTA COOKIN'
 by Bruce Carlson ................................. (3x5) paperback $5.95

## ILLINOIS BOOKS

ILLINOIS COOKIN'
 by Bruce Carlson ................................. (3x5) paperback $5.95
THE VANISHING OUTHOUSE OF ILLINOIS
 by Bruce Carlson ................................................ paperback $9.95
A FIELD GUIDE TO ILLINOIS' CRITTERS
 by Bruce Carlson ................................................ paperback $7.95
YOU KNOW YOU'RE IN ILLINOIS WHEN...
 by Bruce Carlson ................................................ paperback $7.95
Some Pretty Tame, but Kinda Funny Stories About Early
ILLINOIS LADIES-OF-THE-EVENING
 by Bruce Carlson ................................................ paperback $9.95
ILLINOIS' ROADKILL COOKBOOK
 by Bruce Carlson ................................................ paperback $7.95
101 WAYS TO USE A DEAD RIVER FLY
 by Bruce Carlson ................................................ paperback $7.95

HOW TO TALK ILLINOIS
    by Netha Bell ............................................................ paperback $7.95
TALL TALES OF THE MISSISSIPPI RIVER
    by Dan Titus ............................................................ paperback $9.95
TALES OF HACKETT'S CREEK
    by Dan Titus ............................................................ paperback $9.95
UNSOLVED MYSTERIES OF THE MISSISSIPPI
    by Netha Bell ............................................................ paperback $9.95
LOST & BURIED TREASURE OF THE MISSISSIPPI RIVER
    by Netha Bell & Gary Scholl ......................... paperback $9.95
STRANGE FOLKS ALONG THE MISSISSIPPI
    by Pat Wallace ....................................................... paperback $9..95
LET'S US GO DOWN TO THE RIVER 'N...
    by various authors ............................................... paperback $9.95
MISSISSIPPI RIVER PO' FOLK
    by Pat Wallace ....................................................... paperback $9.95
GHOSTS OF THE MISSISSIPPI RIVER (from Keokuk to St. Louis)
    by Bruce Carlson .................................................. paperback $9.95
GHOSTS OF THE MISSISSIPPI RIVER (from Dubuque to Keokuk)
    by Bruce Carlson .................................................. paperback $9.95
MAKIN' DO IN ILLINOIS
    by various authors ............................................... paperback $9.95
MY VERY FIRST
    by various authors ............................................... paperback $9.95
VACANT LOT, SCHOOL YARD & BACK ALLEY GAMES
    by various authors ............................................... paperback $9.95
HOW TO TALK MIDWESTERN
    by Robert Thomas ............................................... paperback $7.95

## INDIANA BOOKS

HOW TO TALK HOOSIER ............................................... paperback $7.95
INDIANA'S ROADKILL COOKBOOK
    by Bruce Carlson .................................................. paperback $7.95
REVENGE OF ROADKILL
    by Bruce Carlson .................................................. paperback $7.95
INDIANA COOKIN'
    by Bruce Carlson .................................... (3x5) paperback $5.95
GHOSTS OF THE OHIO RIVER (from Cincinnati to Louisville)
    by Bruce Carlson .................................................. paperback $9.95
LET'S US GO DOWN TO THE RIVER 'N...
    by various authors ............................................... paperback $9.95

INDIANA PRAIRIE SKIRTS
 by Bev Faaborg & Lois Brinkman .................. paperback $9.95
INDIANA'S VANISHING OUTHOUSE
 by Bruce Carlson ............................................. paperback $9.95
VACANT LOT, SCHOOL YARD & BACK ALLEY GAMES
 by various authors ........................................... paperback $9.95
HOW TO TALK MIDWESTERN
 by Robert Thomas ........................................... paperback $7.95
INDIANA COOKIN'
 by Bruce Carlson .................................. (3x5) paperback $5.95

## IOWA BOOKS

IOWA COOKIN'
 by Bruce Carlson .................................. (3x5) paperback $5.95
IOWA'S ROADKILL COOKBOOK
 by Bruce Carlson ............................................. paperback $7.95
REVENGE OF ROADKILL
 by Bruce Carlson ............................................. paperback $7.95
IOWA'S OLD SCHOOLHOUSES
 by Carole Turner Johnston ............................. paperback $9.95
GHOSTS OF THE AMANA COLONIES
 by Lori Erickson ................................................ paperback $9.95
GHOSTS OF THE IOWA GREAT LAKES
 by Bruce Carlson ............................................. paperback $9.95
GHOSTS OF THE MISSISSIPPI RIVER (from Dubuque to Keokuk)
 by Bruce Carlson ............................................. paperback $9.95
GHOSTS OF THE MISSISSIPPI RIVER (from Minneapolis to Dubuque)
 by Bruce Carlson ............................................. paperback $9.95
GHOSTS OF POLK COUNTY, IOWA
 by Tom Welch ................................................... paperback $9.95
TALES OF HACKETT'S CREEK
 by Dan Titus ...................................................... paperback $9.95
ME 'N WESLEY (stories about the homemade toys that Iowa farm
 children made and played with around the turn of the century)
 by Bruce Carlson ............................................. paperback $9.95
TALL TALES OF THE MISSISSIPPI RIVER
 by Dan Titus ...................................................... paperback $9.95
HOW TO TALK IOWA .................................................. paperback $7.95
UNSOLVED MYSTERIES OF THE MISSISSIPPI
 by Netha Bell ................................................... paperback $9.95
101 WAYS TO USE A DEAD RIVER FLY
 by Bruce Carlson ............................................. paperback $7.95

LET'S US GO DOWN TO THE RIVER 'N...
   by various authors ........................................... paperback $9.95
TRICKS WE PLAYED IN IOWA
   by various authors ........................................... paperback $9.95
IOWA, THE LAND BETWEEN THE VOWELS
   (farm boy stories from the early 1900s)
   by Bruce Carlson ............................................ paperback $9.95
LOST & BURIED TREASURE OF THE MISSISSIPPI RIVER
   by Netha Bell & Gary Scholl .......................... paperback $9.95
Some Pretty Tame, but Kinda Funny Stories About Early
IOWA LADIES-OF-THE-EVENING
   by Bruce Carlson ............................................ paperback $9.95
THE VANISHING OUTHOUSE OF IOWA
   by Bruce Carlson ............................................ paperback $9.95
IOWA'S EARLY HOME REMEDIES
   by 26 students at Wapello Elem. School ..... paperback $9.95
IOWA - A JOURNEY IN A PROMISED LAND
   by Kathy Yoder ............................................. paperback $16.95
LOST & BURIED TREASURE OF THE MISSOURI RIVER
   by Netha Bell ................................................. paperback $9.95
FIELD GUIDE TO IOWA'S CRITTERS
   by Bruce Carlson ............................................ paperback $7.95
OLD IOWA HOUSES, YOUNG LOVES
   by Bruce Carlson ............................................ paperback $9.95
SKUNK RIVER ANTHOLOGY
   by Gene Olson ................................................ paperback $9.95
VACANT LOT, SCHOOL YARD & BACK ALLEY GAMES
   by various authors ........................................... paperback $9.95
HOW TO TALK MIDWESTERN
   by Robert Thomas .......................................... paperback $7.95

## KANSAS BOOKS

HOW TO TALK KANSAS ............................................... paperback $7.95
STOPOVER IN KANSAS
   by Jon McAlpin............................................... paperback $9.95
LET'S US GO DOWN TO THE RIVER 'N...
   by various authors ........................................... paperback $9.95
LOST & BURIED TREASURE OF THE MISSOURI RIVER
   by Netha Bell ................................................... paperback $9.95

101 WAYS TO USE A DEAD RIVER FLY
    by Bruce Carlson ............................................. paperback $7.95
VACANT LOT, SCHOOL YARD & BACK ALLEY GAMES
    by various authors ......................................... paperback $9.95
HOW TO TALK MIDWESTERN
    by Robert Thomas .......................................... paperback $7.95

## KENTUCKY BOOKS

GHOSTS OF THE OHIO RIVER (from Pittsburgh to Cincinnati)
    by Bruce Carlson ............................................ paperback $9.95
GHOSTS OF THE OHIO RIVER (from Cincinnati to Louisville)
    by Bruce Carlson ............................................ paperback $9.95
TALES OF HACKETT'S CREEK
    by Dan Titus ................................................... paperback $9.95
LOST & BURIED TREASURE OF THE MISSISSIPPI RIVER
    by Netha Bell & Gary Scholl .......................... paperback $9.95
LET'S US GO DOWN TO THE RIVER 'N...
    by various authors ......................................... paperback $9.95
UNSOLVED MYSTERIES OF THE MISSISSIPPI
    by Netha Bell ................................................. paperback $9.95
101 WAYS TO USE A DEAD RIVER FLY
    by Bruce Carlson ............................................ paperback $7.95
TALL TALES OF THE MISSISSIPPI RIVER
    by Dan Titus ................................................... paperback $9.95
MY VERY FIRST
    by various authors ......................................... paperback $9.95
VACANT LOT, SCHOOL YARD & BACK ALLEY GAMES
    by various authors ......................................... paperback $9.95

## MICHIGAN BOOKS

MICHIGAN COOKIN'
    by Bruce Carlson .................................. (3x5) paperback $5.95
MICHIGAN'S ROADKILL COOKBOOK
    by Bruce Carlson ............................................ paperback $7.95
MICHIGAN'S VANISHING OUTHOUSE
    by Bruce Carlson ............................................ paperback $9.95

## MINNESOTA BOOKS

MINNESOTA'S ROADKILL COOKBOOK
    by Bruce Carlson .............................................. paperback $7.95
REVENGE OF ROADKILL
    by Bruce Carlson .............................................. paperback $7.95
A FIELD GUIDE TO SMALL MINNESOTA FEMALES
    by Bruce Carlson .............................................. paperback $9.95
GHOSTS OF THE MISSISSIPPI RIVER (from Minneapolis to Dubuque)
    by Bruce Carlson .............................................. paperback $9.95
LAKES COUNTRY COOKBOOK
    by Bruce Carlson .......................................... paperback $11.95
UNSOLVED MYSTERIES OF THE MISSISSIPPI
    by Netha Bell ..................................................... paperback $9.95
TALES OF HACKETT'S CREEK
    by Dan Titus ....................................................... paperback $9.95
GHOSTS OF SOUTHWEST MINNESOTA
    by Ruth Hein ...................................................... paperback $9.95
HOW TO TALK LIKE A MINNESOTA NATIVE ................ Paperback $7.95
MINNESOTA'S VANISHING OUTHOUSE
    . by Bruce Carlson ............................................. paperback $9.95
TALL TALES OF THE MISSISSIPPI RIVER
    by Dan Titus ....................................................... paperback $9.95
Some Pretty Tame, but Kinda Funny Stories About Early
MINNESOTA LADIES-OF-THE-EVENING
    by Bruce Carlson .............................................. paperback $9.95
101 WAYS TO USE A DEAD RIVER FLY ....................... paperback $7.95
LOST & BURIED TREASURE OF THE MISSISSIPPI RIVER
    by Netha Bell & Gary Scholl ........................... paperback $9.95
VACANT LOT, SCHOOL YARD & BACK ALLEY GAMES
    by various authors ............................................ paperback $9.95
HOW TO TALK MIDWESTERN
    by Robert Thomas ............................................ paperback $7.95
MINNESOTA COOKIN'
    by Bruce Carlson ................................... (3x5) paperback $5.95

## MISSOURI BOOKS

MISSOURI COOKIN'
    by Bruce Carlson ................................... (3x5) paperback $5.95
MISSOURI'S ROADKILL COOKBOOK
    by Bruce Carlson .............................................. paperback $7.95

REVENGE OF ROADKILL
    by Bruce Carlson ............................................. paperback $7.95
LET'S US GO DOWN TO THE RIVER 'N...
    by various authors .......................................... paperback $9.95
LAKES COUNTRY COOKBOOK
    by Bruce Carlson ........................................... paperback $11.95
101 WAYS TO USE A DEAD RIVER FLY
    by Bruce Carlson ............................................. paperback $7.95
TALL TALES OF THE MISSISSIPPI RIVER
    by Dan Titus ...................................................... paperback $9.95
TALES OF HACKETT'S CREEK
    by Dan Titus ...................................................... paperback $9.95
STRANGE FOLKS ALONG THE MISSISSIPPI
    by Pat Wallace ................................................ paperback $9.95
LOST & BURIED TREASURE OF THE MISSOURI RIVER
    by Netha Bell ................................................... paperback $9.95
HOW TO TALK MISSOURIAN
    by Bruce Carlson ............................................. paperback $7.95
VACANT LOT, SCHOOL YARD & BACK ALLEY GAMES
    by various authors .......................................... paperback $9.95
HOW TO TALK MIDWESTERN
    by Robert Thomas .......................................... paperback $7.95
UNSOLVED MYSTERIES OF THE MISSISSIPPI
    by Netha Bell ................................................... paperback $9.95
LOST & BURIED TREASURE OF THE MISSISSIPPI RIVER
    by Netha Bell & Gary Scholl ......................... paperback $9.95
MISSISSIPPI RIVER PO' FOLK
    by Pat Wallace ................................................ paperback $9.95
Some Pretty Tame, but Kinda Funny Stories About Early
MISSOURI LADIES-OF-THE-EVENING
    by Bruce Carlson ............................................ paperback $9.95
GUNSHOOTIN', WHISKEY DRINKIN', GIRL CHASIN'
STORIES OUT OF THE OLD MISSOURI TERRITORY
    by Bruce Carlson ............................................ paperback $9.95
THE VANISHING OUTHOUSE OF MISSOURI
    by Bruce Carlson ............................................ paperback $9.95
EARLY MISSOURI HOME REMEDIES
    by various authors .......................................... paperback $9.95
GHOSTS OF THE OZARKS
    by Bruce Carlson ............................................ paperback $9.95

MISSISSIPPI RIVER COOKIN' BOOK
  by Bruce Carlson ................................................. paperback $11.95
MISSOURI'S OLD HOUSES, AND NEW LOVES
  by Bruce Carlson ................................................. paperback $9.95
UNDERGROUND MISSOURI
  by Bruce Carlson ................................................. paperback $9.95

## NEBRASKA BOOKS

LOST & BURIED TREASURE OF THE MISSOURI RIVER
  by Netha Bell ..................................................... paperback $9.95
101 WAYS TO USE A DEAD RIVER FLY
  by Bruce Carlson ................................................. paperback $7.95
LET'S US GO DOWN TO THE RIVER 'N...
  by various authors ............................................... paperback $9.95
HOW TO TALK MIDWESTERN
  by Robert Thomas ................................................. paperback $7.95
VACANT LOT, SCHOOL YARD & BACK ALLEY GAMES
  by various authors ............................................... paperback $9.95

## TENNESSEE BOOKS

TALES OF HACKETT'S CREEK
  by Dan Titus ..................................................... paperback $9.95
TALL TALES OF THE MISSISSIPPI RIVER
  by Dan Titus ..................................................... paperback $9.95
UNSOLVED MYSTERIES OF THE MISSISSIPPI
  By Netha Bell .................................................... paperback $9.95
LOST & BURIED TREASURE OF THE MISSISSIPPI RIVER
  by Netha Bell & Gary Scholl ...................................... paperback $9.95
LET'S US GO DOWN TO THE RIVER 'N...
  by various authors ............................................... paperback $9.95
101 WAYS TO USE A DEAD RIVER FLY
  by Bruce Carlson ................................................. paperback $7.95
VACANT LOT, SCHOOL YARD & BACK ALLEY GAMES
  by various authors ............................................... paperback $9.95

# WISCONSIN BOOKS

HOW TO TALK WISCONSIN .................................................. paperback $7.95
WISCONSIN COOKIN'
    by Bruce Carlson .................................. (3x5) paperback $5.95
WISCONSIN'S ROADKILL COOKBOOK
    by Bruce Carlson ............................................. paperback $7.95
REVENGE OF ROADKILL
    by Bruce Carlson ............................................. paperback $7.95
TALL TALES OF THE MISSISSIPPI RIVER
    by Dan Titus ..................................................... paperback $9.95
LAKES COUNTRY COOKBOOK
    by Bruce Carlson ........................................... paperback $11.95
TALES OF HACKETT'S CREEK
    by Dan Titus ..................................................... paperback $9.95
LET'S US GO DOWN TO THE RIVER 'N...
    by various authors ......................................... paperback $9.95
101 WAYS TO USE A DEAD RIVER FLY
    by Bruce Carlson ............................................. paperback $7.95
UNSOLVED MYSTERIES OF THE MISSISSIPPI
    by Netha Bell .................................................. paperback $9.95
LOST & BURIED TREASURE OF THE MISSISSIPPI RIVER
    by Netha Bell & Gary Scholl .......................... paperback $9.95
GHOSTS OF THE MISSISSIPPI RIVER (from Dubuque to Keokuk)
    by Bruce Carlson ............................................. paperback $9.95
HOW TO TALK MIDWESTERN
    by Robert Thomas ........................................... paperback $7.95
VACANT LOT, SCHOOL YARD & BACK ALLEY GAMES
    by various authors ......................................... paperback $9.95
MY VERY FIRST
    by various authors ......................................... paperback $9.95
EARLY WISCONSIN HOME REMEDIES
    by various authors ......................................... paperback $9.95
GHOSTS OF THE MISSISSIPPI RIVER (from Minneapolis to Dubuque)
    by Bruce Carlson ............................................. paperback $9.95
THE VANISHING OUTHOUSE OF WISCONSIN
    by Bruce Carlson ............................................. paperback $9.95
GHOSTS OF DOOR COUNTY, WISCONSIN
    by Geri Rider ................................................... paperback $9.95
Some Pretty Tame, but Kinda Funny Stores About Early
WISCONSIN LADIES-OF-THE-EVENING
    by Bruce Carlson ............................................. paperback $9.95

## MIDWESTERN BOOKS

A FIELD GUIDE TO THE MIDWEST'S WORST RESTAURANTS
  by Bruce Carlson ............................................. paperback $5.95
THE MOTORIST'S FIELD GUIDE TO MIDWESTERN FARM
EQUIPMENT (misguided information as only a city slicker can give it)
  by Bruce Carlson ............................................. paperback $5.95
VACANT LOT, SCHOOL YARD & BACK ALLEY GAMES
OF THE MIDWEST YEARS AGO
  by various authors .......................................... paperback $9.95
MIDWEST SMALL TOWN COOKING
  by Bruce Carlson .................................. (3x5) paperback $5.95
HITCHHIKING THE UPPER MIDWEST
  by Bruce Carlson ............................................ paperback $7.95
101 WAYS FOR MIDWESTERNERS TO "DO IN" THEIR
NEIGHBOR'S PESKY DOG WITHOUT GETTING CAUGHT
  by Bruce Carlson ............................................ paperback $5.95

## RIVER BOOKS

ON THE SHOULDERS OF A GIANT
  by M. Cody and D. Walker ........................... paperback $9.95
SKUNK RIVER ANTHOLOGY
  by Gene "Will" Olson ..................................... paperback $9.95
JACK KING vs. DETECTIVE MACKENZIE
  by Netha Bell ................................................... paperback $9.95
LOST & BURIED TREASURES ALONG THE MISSISSIPPI
  by Netha Bell & Gary Scholl ......................... paperback $9.95
MISSISSIPPI RIVER PO' FOLK
  by Pat Wallace ............................................... paperback $9.95
STRANGE FOLKS ALONG THE MISSISSIPPI
  by Pat Wallace ............................................... paperback $9.95
GHOSTS OF THE OHIO RIVER (from Pittsburgh to Cincinnati)
  by Bruce Carlson ............................................ paperback $9.95
GHOSTS OF THE OHIO RIVER (from Cincinnati to Louisville)
  by Bruce Carlson ............................................ paperback $9.95
GHOSTS OF THE MISSISSIPPI RIVER (Minneapolis to Dubuque)
  by Bruce Carlson ............................................ paperback $9.95
GHOSTS OF THE MISSISSIPPI RIVER (Dubuque to Keokuk)
  by Bruce Carlson ............................................ paperback $9.95
TALL TALES OF THE MISSISSIPPI RIVER
  by Dan Titus .................................................... paperback $9.95

TALL TALES OF THE MISSOURI RIVER
    by Dan Titus ....................................................... paperback $9.95
RIVER SHARKS & SHENANIGANS
    (tales of riverboat gambling of years ago)
    by Netha Bell .................................................. paperback $9.95
UNSOLVED MYSTERIES OF THE MISSISSIPPI
    by Netha Bell .................................................. paperback $9.95
TALES OF HACKETT'S CREEK (1940s Mississippi River kids)
    by Dan Titus ....................................................... paperback $9.95
101 WAYS TO USE A DEAD RIVER FLY
    by Bruce Carlson .............................................. paperback $7.95
LET'S US GO DOWN TO THE RIVER 'N...
    by various authors ........................................... paperback $9.95
LOST & BURIED TREASURE OF THE MISSOURI
    by Netha Bell .................................................. paperback $9.95

## COOKBOOKS

ROARING 20'S COOKBOOK
    by Bruce Carlson ........................................... paperback $11.95
DEPRESSION COOKBOOK
    by Bruce Carlson ........................................... paperback $11.95
LAKES COUNTRY COOKBOOK
    by Bruce Carlson ........................................... paperback $11.95
A COOKBOOK FOR THEM WHAT AIN'T DONE A LOT OF COOKIN'
    by Bruce Carlson ........................................... paperback $11.95
FLAT-OUT DIRT-CHEAP COOKIN' COOKBOOK
    by Bruce Carlson ........................................... paperback $11.95
APHRODISIAC COOKING
    by Bruce Carlson ........................................... paperback $11.95
WILD CRITTER COOKBOOK
    by Bruce Carlson ........................................... paperback $11.95
I GOT FUNNIER-THINGS-TO-DO-THAN-COOKIN' COOKBOOK
    by Louise Lum ................................................ paperback $11.95
MISSISSIPPI RIVER COOKIN' BOOK
    by Bruce Carlson ........................................... paperback $11.95
HUNTING IN THE NUDE COOKBOOK
    by Bruce Carlson ............................................. paperback $9.95
DAKOTA COOKIN'
    by Bruce Carlson .................................... (3x5) paperback $5.95
IOWA COOKIN'
    by Bruce Carlson .................................... (3x5) paperback $5.95

MICHIGAN COOKIN'
by Bruce Carlson .................................... (3x5) paperback $5.95
MINNESOTA COOKIN'
by Bruce Carlson .................................... (3x5) paperback $5.95
MISSOURI COOKIN'
by Bruce Carlson .................................... (3x5) paperback $5.95
ILLINOIS COOKIN'
by Bruce Carlson .................................... (3x5) paperback $5.95
WISCONSIN COOKIN'
by Bruce Carlson .................................... (3x5) paperback $5.95
HILL COUNTRY COOKIN'
by Bruce Carlson .................................... (3x5) paperback $5.95
MIDWEST SMALL TOWN COOKIN'
by Bruce Carlson .................................... (3x5) paperback $5.95
APHRODISIAC COOKIN'
by Bruce Carlson .................................... (3x5) paperback $5.95
PREGNANT LADY COOKIN'
by Bruce Carlson .................................... (3x5) paperback $5.95
GOOD COOKIN' FROM THE PLAIN PEOPLE
by Bruce Carlson .................................... (3x5) PAPERBACK $5.95
WORKING GIRL COOKING
by Bruce Carlson .................................... (3x5) paperback $5.95
COOKING FOR ONE
by Barb Layton ............................................. paperback $11.95
SUPER SIMPLE COOKING
by Barb Layton ........................................ (3x5) paperback $5.95
OFF TO COLLEGE COOKBOOK
by Barb Layton ........................................ (3x5) paperback $5.95
COOKING WITH THINGS THAT GO SPLASH
by Bruce Carlson .................................... (3x5) paperback $5.95
COOKING WITH THINGS THAT GO MOO
by Bruce Carlson .................................... (3x5) paperback $5.95
COOKING WITH SPIRITS
by Bruce Carlson .................................... (3x5) paperback $5.95
INDIAN COOKING COOKBOOK
by Bruce Carlson ............................................ paperback $9.95
DIAL-A-DREAM COOKBOOK
by Bruce Carlson ........................................... paperback $11.95
HORMONE HELPER COOKBOOK .......................... paperback $11.95
INDIANA COOKIN'
by Bruce Carlson .................................... (3x5) paperback $5.95

## MISCELLANEOUS BOOKS

DEAR TABBY (letters to and from a feline advice columnist)
 by Bruce Carlson ............................................. paperback $5.95
HOW TO BEHAVE (etiquette advice for non-traditional
and awkward circumstances such as attending dogfights,
what to do when your blind date turns out to be your spouse, etc.)
 by Bruce Carlson ............................................. paperback $5.95
REVENGE OF THE ROADKILL
 by Bruce Carlson ............................................. paperback $7.95

# SUBSTITUTIONS

| FOR | YOU CAN USE... |
|---|---|
| 1 T. cornstarch | 2 T. flour OR 1½ T. quick cooking tapioca |
| 1 C. cake flour | 1 C. less 2 T. all-purpose flour |
| 1 C. all-purpose flour | 1 C. plus 2 T. cake flour |
| 1 sq. chocolate | 3 T. cocoa & 1 T. fat |
| 1 C. melted shortening | 1 C. salad oil (may not be substituted for solid shortening) |
| 1 C. milk | ½ C. evaporated milk & ½ C. water |
| 1 C. sour milk or buttermilk | 1 T. lemon juice or vinegar & enough sweet milk to measure 1 C. |
| 1 C. heavy cream | ⅔ C. milk & ⅓ C. butter |
| Sweetened condensed milk | No substitution |
| 1 egg | 2 T. dried whole egg & 2 T. water |
| 1 tsp. baking powder | ¼ tsp. baking soda & 1 tsp. cream of tartar OR ¼ tsp. baking soda & ½ C. sour milk, buttermilk or molasses; reduce other liquid ½ C. |
| 1 C. sugar | 1 C. honey; reduce other liquid ¼ C.; reduce baking temperature by 25° |
| 1 C. miniature marshmallows | About 10 large marshmallows (cut-up) |
| 1 medium onion (2½-inch diameter) | 2 T. instant minced onion OR 1 tsp. onion powder OR 2 T. onion salt; reduce salt 1 tsp. |
| 1 garlic clove | 1/8 tsp. garlic powder OR ¼ tsp. garlic salt; reduce salt 1/8 tsp. |
| 1 T fresh herbs | 1 tsp. dried herbs OR ¼ tsp. powdered herbs OR ½ tsp. herb salt; reduce salt ¼ tsp. |

## Protein Content and Caloric Value of Foods for Your Diet

| Food | Oz. | Approximate Measure | Protein | Calories |
|---|---|---|---|---|
| **Lamb** | | | | |
| Chops | | | | |
|   Loin or rib | 4 | 1 loin or 2 rib 1-inch thick | 17.9 | 421 |
|   Shoulder | 4 | Piece 4x3x5/8-inch | 18.7 | 348 |
| Roasts | | | | |
|   Leg | 4 | Slice 4x3x½-inch | 21.6 | 276 |
|   Shoulder | 4 | Slice 5x3x½-inch | 18.7 | 348 |
| **Pork, fresh** | | | | |
| Chops and steaks | | | | |
|   Leg (ham) | 4 | Piece 3½x3x½-inch | 18.2 | 408 |
|   Loin | 4 | Chop ¾-inch thick | 19.7 | 349 |
|   Shoulder | 4 | Piece 4½x3½x3/8-inch | 16.1 | 464 |
| Roasts | | | | |
|   Boston butt | 4 | Slice 4½x3½x3/8-inch | 19.9 | 327 |
|   Loin | 4 | Slice ¾-inch thick | 19.7 | 349 |
|   Tenderloin | 4 | 2 pieces 1-inch dia.x3-inches long | 23.9 | 172 |
| **Pork, cured** | | | | |
| Bacon, Canadian style | 1 | Slice 2¼-inch diameter by 3/16-inch thick | 6.6 | 68 |
| Ham (boiled) | 2 | Slice 4¼x4x1/8-inch | 10.6 | 147 |
| **Veal** | | | | |
| Chops | | | | |
|   Loin | 4 | Chop 5/8-inch thick | 23.0 | 211 |
|   Rib | 4 | Chop ¾-inch thick | 22.6 | 241 |
| Roasts | | | | |
|   Leg | 4 | Slice 4x2½x½-inch | 22.9 | 223 |
|   Loin | 4 | Slice 4x2½x½-inch | 23.0 | 211 |
|   Rib | 4 | Slice 4x2½x½-inch | 22.6 | 241 |
|   Shoulder | 4 | Slice 5x3x½-inch | 23.3 | 202 |
| Steaks | | | | |
|   Cutlet (round) | 4 | Piece 4x2½x½-inch | 23.4 | 191 |
|   Shoulder | 4 | Piece 5x3x½-inch | 23.3 | 202 |
|   Sirloin | 4 | Piece 4x2½x½-inch | 23.0 | 211 |
| Stew (breast) | 4 | 4 pieces 2½x1x1-inch | 22.0 | 271 |
| **Variety Meats** | | | | |
| Brains (beef) | 4 | 2 pieces 2½x1½x1-inch | 12.6 | 152 |
| Heart (avg.) | 4 | ⅓ ht. 3-inch dia. x 3½-inch long | 19.7 | 157 |
| Kidney (avg.) | 4 | 3 slices 3¼x2½x¼-inch | 20.0 | 161 |
| **Liver** | | | | |
| Beef | 3 | 2 slices 3x2½x3/8-inch | 17.7 | 119 |
| Lamb | 3 | 2 slices 3½x2x3/8-inch | 18.9 | 118 |
| Pork | 3 | 2 slices 3½x2x3/8-inch | 17.7 | 116 |
| Veal | 3 | 2 slices 3x2½x3/8-inch | 17.1 | 122 |
| Sweetbread | 4 | Piece 4x3x¾-inch | 18.2 | 216 |
| Tongue | 3 | 3 slices 3x2x¼-inch | 15.7 | 191 |

# Protein Content and Caloric Value of Foods for Your Diet

| Food | Oz. | Approximate Measure | Protein | Calories |
|---|---|---|---|---|
| **Sausages and Cooked Specialties** | | | | |
| Bologna | 1 | Slice 4½ dia. x ½-inch thick | 4.4 | 65 |
| Frankfurter | 2 | 2 5½-inch long x ¾-inch dia. | 9.1 | 121 |
| Liver sausage | 1 | Slice 3-inch dia. x ¼-inch thick | 5.0 | 77 |
| Luncheon meat | 1 | Slice 4x3½x1/8-inch | 4.6 | 81 |
| Vienna sausage | 1 | 2 pieces 2-inch long x ¾-inch diameter | 5.8 | 76 |
| **Poultry** | | | | |
| Chicken | | | | |
| Liver | 3 | 4 avg. | 19.9 | 122 |
| Roast | | | | |
|   Breast | 3 | ½ breast | 21.0 | 110 |
|   Leg | 2½ | 1 avg. | 14.7 | 88 |
|   Thigh | 2½ | 1 avg. | 15.8 | 95 |
|   Wing | 1 | 1 avg. | 7.0 | 37 |
| Stewed | | | | |
|   Dark meat | 3½ | ½ cup (diced) | 23.1 | 139 |
|   Light meat | 3 | ½ cup (diced) | 20.3 | 106 |
| **Turkey** | | | | |
| Roast | | | | |
|   Dark meat | 3½ | Slice 4x3x½-inch | 23.2 | 177 |
|   Light meat | 3½ | Slice 4x3x½-inch | 24.5 | 139 |
| **Fish** | | | | |
| Bass | 4 | 1 small fish | 27.3 | 113 |
| Clams | 3½ | 5 medium | 12.8 | 77 |
| Cod | 3½ | Piece 4x2¼x¾-inch | 16.5 | 70 |
| Crab (canned) | 3 | ⅔ cup | 16.1 | 94 |
| Finnan haddie | 3½ | ¾ cup | 23.2 | 96 |
| Flounder | 3½ | Piece 4x3x3/8-inch | 19.0 | 79 |
| Haddock | 3½ | Piece 3½x3x¾-inch | 17.2 | 72 |
| Halibut | 4 | Piece 4x3x½-inch | 20.4 | 133 |
| Herring, fresh | 4 | 1 fish 7-inches long | 22.8 | 163 |
| **Lobster** | | | | |
| Canned | 3 | ½ cup | 15.6 | 74 |
| Fresh | 2½ | 1 avg. | 12.2 | 63 |
| Mackerel | 2½ | ¼ fish 7-inches long | 14.3 | 119 |
| Oysters | 3½ | 5 medium | 6.0 | 50 |
| Perch | 4 | 2 fish 4½-inches long | 23.4 | 102 |
| Salmon | | | | |
|   Canned | 3½ | ⅔ cup | 24.7 | 203 |
|   Fresh | 3 | Piece 2½x2½x7/8-inch | 15.7 | 196 |
| Shrimp (canned) | 2 | 3/8 cup or 12 pieces 1-inch dia. | 10.7 | 49 |
| Trout | 3 | Piece 6-inches long | 16.1 | 80 |
| White fish | 4 | Piece 3¼-inchx3x½-inch | 25.2 | 165 |
| **Milk and Dairy Products** | | | | |
| Butter | ⅓ | | .1 | 73 |
| Cheese, | | | | |
|   cottage | 2 | ¼ cup | 9.6 | 51 |
| Cream, coffee | ½ | 1 T. | .4 | 29 |

## Protein Content and Caloric Value of Foods for Your Diet

| Food | Approx. Weight (Oz.) | Approximate Measure (Gm.) | Protein | Calories |
|---|---|---|---|---|
| **Milk** | | | | |
| Buttermilk | 7 | 1 glass | 7.0 | 72 |
| Evaporated | 4 | ½ cup | 8.4 | 167 |
| Skim | 7 | 1 glass | 7.0 | 72 |
| Whole | 7 | 1 glass | 7.0 | 138 |
| **Eggs** | 1⅔ | 1 medium | 6.4 | 79 |
| **Potatoes** | | | | |
| White | 2 | 1 small 2½-inch long x 2-inch dia. | 1.2 | 51 |
| **Vegetables** | | | | |
| Artichokes | 3½ | ½ large | 2.9 | 63 |
| Asparagus | 3½ | 7 stalks 6-inches long | 2.3 | 27 |
| Beans, string | 3½ | ⅔ cup | 2.4 | 42 |
| Beet greens | 3½ | ½ cup | 2.0 | 33 |
| Beets | 3½ | ⅔ cup or 2 1¾-inch dia. | 1.6 | 46 |
| Broccoli | 3½ | 2 stalks 5-inches long | 3.3 | 37 |
| Brussels sprts. | 3½ | ⅔ cup | 4.4 | 58 |
| Cabbage | 3½ | 1/5 head 4½-inch dia. | 1.4 | 29 |
| Carrots | 3½ | 2 carrots 5-inch long | 1.2 | 45 |
| Cauliflower | 3½ | ⅔ cup | 2.4 | 31 |
| Celery | ½ | Piece 8½-inch long or 2 hts. | .2 | 3 |
| Chard, Swiss | 3½ | ½ cup | 1.4 | 25 |
| Chicory | 1 | 10 small leaves | .4 | 7 |
| Cucumbers | 2 | 8 slices 1/8-inch thick | .4 | 7 |
| Eggplant | 2 | Slices 3½-inch dia x 3/8-inch thick | .7 | 17 |
| Endive, French | 2 | 2 stalks | .8 | 11 |
| Green pepper | ½ | ½ cup or piece 4x1¾-inch | .2 | 4 |
| Kohlrabi | 3½ | ⅔ cup (diced) | 2.1 | 36 |
| Lettuce | | | | |
| Head | 3½ | ¼ head 4-inch diameter | 1.2 | 18 |
| | ½ | 1 leaf | .2 | 3 |
| Leaf | ½ | 2 leaves | .1 | 2 |
| Mushrooms | 3½ | 5 caps 2¼-inch dia. | 2.6 | 15 |
| Okra | 2 | 5 pods | 1.0 | 21 |
| Onions | | | | |
| Dried | 3 | 1 onion 2-inch dia. | 1.2 | 42 |
| Green | ½ | 3 medium | .2 | 7 |
| Parsley | | 2 sprigs | .1 | 1 |
| Pumpkin | 3½ | ½ cup | 1.2 | 36 |
| Radishes | 1 | 3 radishes 1-inch dia. | .4 | 7 |
| Rutabagas | 3½ | ½ cup | 1.1 | 41 |
| Sauerkraut | 3½ | ⅔ cup | 1.1 | 18 |
| Spinach | 3½ | ¾ cup | 2.3 | 25 |
| Squash | | | | |
| Summer | 3½ | ½ cup | .6 | 19 |
| Winter | 3½ | ½ cup | 1.5 | 44 |
| Tomatoes | | | | |
| Canned | 3½ | ½ cup | 1.2 | 25 |
| Fresh | 3½ | 1 tomato 2-inch dia. | 1.0 | 23 |
| Juice, canned | 4 | ½ cup | 1.2 | 28 |
| Turnip greens | 3½ | ½ cup | 2.9 | 37 |

## Protein Content and Caloric Value of Foods for Your Diet

| Food | Oz. | Approximate Measure | Protein | Calories |
|---|---|---|---|---|
| Turnips | | | | |
|   White | 3½ | ⅔ cup | 1.1 | 35 |
|   Yellow (see | rutabagas) | | | |
| Pickles | | | | |
|   Olives | | | | |
|     Green | 1/6 | 1 medium | .1 | 7 |
|     Ripe | ½ | 1 large | .2 | 23 |
|   Pickles | | | | |
|     Dill | 2 | ½ pickle 5-inches long x1½-inch diameter | .3 | 7 |
|     Sweet | ½ | 1 pickle 2½-inches long x¾-inch diameter | .2 | 21 |

### BREAD AND CEREAL PRODUCTS

| Food | Oz. | Approximate Measure | Protein | Calories |
|---|---|---|---|---|
| Cereals | | | | |
|   Bran, whole | ⅔ | ⅓ cup | 2.5 | 67 |
|   Cornflakes | ½ | ⅔ cup | 1.3 | 56 |
|   Farina, enriched | ⅔ | ½ cup (sc. 2 T. dry) | 2.3 | 71 |
|   Oatmeal | ⅔ | ½ cup (¼ cup dry) | 3.1 | 77 |
| Rice | | | | |
|   Puffed | ⅓ | ¾ cup | .7 | 36 |
|   White | 1 | ⅔ cup (2 T. dry) | 2.3 | 105 |
| Wheat | | | | |
|   Flakes | ⅔ | ¾ cup | 2.4 | 74 |
|   Puffed | ⅓ | ¾ cup | 1.2 | 37 |
|   Shredded | 1 | 1 biscuit | 2.9 | 103 |
| Breads | | | | |
|   Rye | ⅔ | Slice 4x3½x½-inch | 1.2 | 50 |
|   Wheat | | | | |
|     Melba toast | 1/6 | Slice 3x2x¼-inch | .6 | 19 |
|     White, enrch | ⅔ | 1 slice (commercial) thin | 1.6 | 50 |
|     Whole wheat | ⅔ | 1 slice (commercial) thin | 1.8 | 50 |
| Crackers | | | | |
|   Graham | ½ | 1 cracker 3-inch square | 1.0 | 54 |
|   Saltine | ½ | 1 cracker 2-inch square | .4 | 17 |
|   Soda | 1/5 | 1 cracker 2¾x2½-inch | .6 | 25 |
|   Zwieback | ¼ | 1 piece 3¼x1¼x½-inch | .9 | 33 |

### Beverages

| Food | Oz. | Approximate Measure | Protein | Calories |
|---|---|---|---|---|
| Carbonated | 6 | 1 small bottle | | 82 |
| Coffee, black | | | 0 | 0 |
| Tea, plain | | | 0 | 0 |

### Fruits

| Food | Oz. | Approximate Measure | Protein | Calories |
|---|---|---|---|---|
| Apples | 3½ | 1 apple 2¼-inch diameter | .3 | 65 |
| Apricots | 1 | 1 medium | .4 | 20 |
| Blackberries | 3½ | ¾ cup | 1.2 | 62 |
| Blueberries | 3½ | ⅔ cup | .6 | 68 |
| Cantaloupe | 4 | ¼ melon 5-inch diameter | .8 | 29 |
| Cherries, sweet | 3½ | 15 cherries 7/8-inch diameter | 1.2 | 87 |
| Grapefruit | 3½ | ½ medium 3 5/8-inch dia. | .5 | 44 |
| Grapes | | | | |
|   Concord | 3½ | 34 avg. | 1.4 | 78 |
|   Green seedless | 3½ | 40 small | .8 | 74 |
|   Malaga or Tokay | 3½ | 21 avg. | .8 | 74 |

## Protein Content and Caloric Value of Foods for Your Diet

| Food | | Serving | Protein | Calories |
|---|---|---|---|---|
| Honeydew melon | 4 | 1½-inch slice, 7-inch melon | .9 | 48 |
| Oranges | 3½ | ½ orange 4-inch diameter | .9 | 52 |
| Peaches | 3½ | 1 medium | .5 | 51 |
| Pears | 3½ | 1 small | .7 | 70 |
| Pineapple | 3½ | 1 slice 4-inch diameter x ½-inch thick | .4 | 58 |
| Plums | 2½ | 1 plum 1¾-inch dia. | .5 | 39 |
| Raspberries | 3 | ⅔ cup | 1.1 | 64 |
| Strawberries | 3½ | 10 strawberries 1-inch dia. | .8 | 41 |
| Watermelon | 5 | ½ slice 6-inch dia. x ¾-inch thick | .8 | 51 |
| **FRUIT JUICES** | | | | |
| Grapefruit, canned | 4 | ½ cup | .6 | 49 |
| Orange | 4 | ½ cup | .7 | 66 |
| Pineapple canned | 4 | ½ cup | .4 | 65 |
| Tomato (see | vegetables) | | | |

# HOW MANY DROPS IN A "DASH"?

## Here, a cook's guide to the most-often-called-for food measures and equivalents

How many cups of berries in a pint? How many slices of bread make a half cup of crumbs? For two tablespoons of orange peel, will you need more than one orange? You'll find the answers to these questions and lots more in this handy kitchen chart.

## EQUIVALENT MEASURES

| | |
|---|---|
| Dash | 2 to 3 drops or less than 1/8 teaspoon |
| 1 tablespoon | 3 teaspoons |
| ¼ cup | 4 tablespoons |
| ⅓ cup | 5 tablespoons plus 1 teaspoon |
| ½ cup | 8 tablespoons |
| 1 cup | 16 tablespoons |
| 1 pint | 2 cups |
| 1 quart | 4 cups |
| 1 gallon | 4 quarts |
| 1 peck | 8 quarts |
| 1 bushel | 4 pecks |
| 1 pound | 16 ounces |

## FOOD EQUIVALENTS

| | |
|---|---|
| **Apples** 1 pound | 3 medium (3 cups sliced) |
| **Bananas** 1 pound | 3 medium (1⅓ cups mashed) |
| **Berries** 1 pint | 1¾ cups |
| **Bread** 1 pound loaf | 14 to 20 slices |
| **Bread crumbs, fresh** 1 slice bread with crust | ½ cup bread crumbs |
| **Broth, chicken or beef** 1 cup | 1 bouillon cube or 1 envelope bouillon or 1 teaspoon instant bouillon dissolved in 1 cup boiling water |
| **Butter or margarine** ¼ pound stick | ½ cup |
| **Cheese** ¼ pound | 1 cup, shredded |
| **Cheese, cottage** 8 ounces | 1 cup |
| **Cheese, cream** 3 ounces | 6 tablespoons |
| **Chocolate, unsweetened** 1 ounce | 1 square |
| **Chocolate, semi-sweet pieces** 6 ounce package | 1 cup |

# QUANTITIES TO SERVE 100 PEOPLE

| | |
|---|---|
| Coffee | —3 lbs. |
| Loaf Sugar | —3 lbs. |
| Cream | —3 qts. |
| Whipping Cream | —4 pts. |
| Milk | —6 gallons |
| Fruit | —2½ gallons |
| Fruit Juice | —4 (No. 10 ea.) cans (26 lbs.) |
| Tomato Juice | —4 (No. 10 ea.) cans (26 lbs.) |
| Soup | —5 gallons |
| Oysters | —18 qts. |
| Weiners | —25 lbs. |
| Meatloaf | —24 lbs. |
| Ham | —40 lbs. |
| Beef | —40 lbs. |
| Roast Pork | —40 lbs. |
| Hamburger | —30-36 lbs. |
| Chicken For Chicken Pie | —40 lbs. |
| Potatoes | —35 lbs. |
| Scalloped Potatoes | —5 gallon |
| Vegetables | —4 (No. 10 ea.) cans (26 lbs.) |
| Baked Beans | —5 gallon |
| Beets | —30 lbs. |
| Cauliflower | —18 lbs. |
| Cabbage For Slaw | —20 lbs. |
| Carrots | —33 lbs. |
| Bread | —10 loaves |
| Rolls | —200 |
| Butter | —3 lbs. |
| Potato Salad | —12 qts. |
| Fruit Salad | —20 qts. |
| Vegetable Salad | —20 qts. |
| Lettuce | —20 heads |
| Salad Dressing | —3 qts. |
| Pies | —18 |
| Cakes | —8 |
| Ice Cream | —4 gallons |
| Cheese | —3 lbs. |
| Olives | —1¾ lbs. |
| Pickles | —2 qts. |
| Nuts | —3 lbs. sorted |

To Serve 50 People, Divide by 2
To Serve 25 People, Divide by 4

# THE KITCHEN
## General Household Hints

## SALT

If stew is too salty, add raw cut potatoes and discard once they have cooked and absorbed the salt. Another remedy is to add a teaspoon each of cider vinegar and sugar. Or, simply add sugar.

If soup or stew is too sweet, add salt. For a main dish or vegetable, add a teaspoon of cider vinegar.

## GRAVY

For pale gravy, color with a few drops of Kitchen Bouquet. Or to avoid the problem in the first place, brown the flour well before adding the liquid. This also helps prevent lumpy gravy.

To make gravy smooth, keep a jar with a mixture of equal parts of flour and cornstarch. Put 3-4 T. of this mixture in another jar and add some water. Shake, and in a few minutes you will have a smooth paste for gravy.

To remedy greasy gravy, add a small amount of baking soda.

For quick thickener for gravies, add some instant potatoes to your gravy and it will thicken beautifully.

## VEGETABLES

If fresh vegetables are wilted or blemished, pick off the brown edges. Sprinkle with cool water, wrap in towel and refrigerate for an hour or so.

Perk up soggy lettuce by adding lemon juice to a bowl of cold water and soak for an hour in the refrigerator.

Lettuce and celery will crisp up fast if you place it in a pan of cold water and add a few sliced potatoes.

If vegetables are overdone, put the pot in a pan of cold water. Let it stand from 15 minutes to ½ hour without scraping pan.

By lining the crisper section of your refrigerator with newspaper and wrapping vegetables with it, moisture will be absorbed and your vegetables will stay fresher longer.

## EGGS

If you shake the egg and you hear a rattle, you can be sure it's stale. A really fresh egg will float and a stale one will sink.

If you are making deviled eggs and want to slice it perfectly, dip the knife in water first. The slice will be smooth with no yolk sticking to the knife.

The white of an egg is easiest to beat when it's at room temperature. So leave it out of the refrigerator about ½ hour before using it.

To make light and fluffy scrambled eggs, add a little water while beating the eggs.

Add vinegar to the water while boiling eggs. Vinegar helps to seal the egg, since it acts on the calcium in the shell.

To make quick-diced eggs, take your potato masher and go to work on a boiled egg.

If you wrap each egg in aluminum foil before boiling it, the shell won't crack when it's boiling.

To make those eggs go further when making scrambled eggs for a crowd, add a pinch of baking powder and 2 tsp. of water per egg.

A great trick for peeling eggs the easy way - when they are finished boiling, turn off the heat and just let them sit in the pan with the lid on for about 5 minutes. Steam will build up under the shell and they will just fall away.

Or, quickly rinse hot hard-boiled eggs in cold water, and the shells will be easier to remove.

When you have saved a lot of egg yolks from previous recipes; use them in place of whole eggs for baking or thickening. Just add 2 yolks for every whole egg.

Fresh or hard-boiled? Spin the egg. If it wobbles, it is raw - if it spins easily, it's hard-boiled.

Add a few drops of vinegar to the water when poaching an egg to keep it from running all over the pan.

Add 1 T. of water per egg white to increase the quantity of beaten egg white when making meringue.

Try adding eggshells to coffee after it has perked, for a better flavor.

## POTATOES

Overcooked potatoes can become soggy when the milk is added. Sprinkle with dry powdered milk for the fluffiest mashed potatoes ever.

To hurry up baked potatoes, boil in salted water for 10 minutes, then place in a very hot oven. Or, cut potatoes in half and place them face down on a baking sheet in the oven to make the baking time shorter.

When making potato pancakes, add a little sour cream to keep potatoes from discoloring.

Save some of the water in which the potatoes were boiled - add to some powdered milk and use when mashing. This restores some of the nutrients that were lost in the cooking process.

Use a couple of tablespoons of cream cheese in place of butter for your potatoes; try using sour cream instead of milk when mashing.

## ONIONS

To avoid tears when peeling onions, peel them under cold water or refrigerate before chopping.

For sandwiches to go in lunchboxes, sprinkle with dried onion. They will have turned into crisp pieces by lunchtime.

Peel and quarter onions. Place one layer deep in a pan and freeze. Quickly pack in bags or containers while frozen. Use as needed, chopping onions while frozen, with a sharp knife.

## TOMATOES

Keep tomatoes in storage with stems pointed downward and they will retain their freshness longer.

Sunlight doesn't ripen tomatoes. It's the warmth that makes them ripen. So find a warm spot near the stove or dishwasher where they can get a little heat.

Save the juice from canned tomatoes in ice cube trays. When frozen, store in plastic bags in freezer for cooking use or for tomato drinks.

To improve the flavor of inexpensive tomato juice, pour a 46-ounce can of it into a refrigerator jar and add one chopped green onion and a cut-up stalk of celery.

## ROCK-HARD BROWN SUGAR
Add a slice of soft bread to the package of brown sugar, close the bag tightly, and in a few hours the sugar will be soft again. If you need it in a hurry, simply grate the amount called for with a hand grater. Or, put brown sugar and a cup of water (do not add to the sugar, set it alongside of it) in a covered pan. Place in the oven (low heat) for awhile. Or, buy liquid brown sugar.

## THAWING FROZEN MEAT
Seal the meat in a plastic bag and place in a bowl of very warm water. Or, put in a bag and let cold water run over it for an hour or so.

## CAKED OR CLOGGED SALT
Tightly wrap a piece of aluminum foil around the salt shaker. This will keep the dampness out of the salt. To prevent clogging, keep 5 to 10 grains of rice inside your shaker.

## SOGGY POTATO CHIPS, CEREAL AND CRACKERS
If potato chips lose their freshness, place under broiler for a few moments. Care must be taken not to brown them. You can crisp soggy cereal and crackers by putting them on a cookie sheet and heating for a few minutes in the oven.

## PANCAKE SYRUP
To make an inexpensive syrup for pancakes, save small amounts of leftover jams and jellies in a jar. Or, fruit-flavored syrup can be made by adding 2 C. sugar to 1 C. of any kind of fruit juice and cooking until it boils.

## EASY TOPPING
A good topping for gingerbread, coffeecake, etc., can easily be made by freezing the syrup from canned fruit and adding 1 T. of butter and 1 T. of lemon juice to 2 C. of syrup. Heat until bubbly, and thicken with 2 T. of flour.

## TASTY CHEESE SANDWICHES
Toast cheese sandwiches in a frying pan lightly greased with bacon fat for a delightful new flavor.

## HURRY-UP HAMBURGERS
Poke a hole in the middle of the patties while shaping them. The burgers will cook faster and the holes will disappear when done.

## SHRINKLESS LINKS
Boil sausage links for about 8 minutes before frying and they will shrink less and not break at all. Or, you can roll them lightly in flour before frying.

## FROZEN BREAD
Put frozen bread loaves in a clean brown paper bag and place for 5 minutes in a 325° oven to thaw completely.

## REMOVING THE CORN SILK
Dampen a paper towel or terry cloth and brush downward on the cob of corn. Every strand should come off.

## NUTS
To quickly crack open a large amount of nuts, put in a bag and gently hammer until they are cracked open. Then remove nutmeats with a pick.

If nuts are stale, place them in the oven at 250° and leave them there for 5 to 10 minutes. The heat will revive them.

## PREVENTING BOIL-OVERS
Add a lump of butter or a few teaspoons of cooking oil to the water. Rice, noodles or spaghetti will not boil over or stick together.

## SOFTENING BUTTER
Soften butter quickly by grating it. Or heat a small pan and place it upside-down over the butter dish for several minutes. Or place in the microwave for a few seconds.

## MEASURING STICKY LIQUIDS
Before measuring honey or syrup, oil the cup with cooking oil and rinse in hot water.

## SCALDED MILK
Add a bit of sugar (without stirring) to milk to prevent it from scorching.

Rinse the pan in cold water before scalding milk, and it will be much easier to clean.

## TENDERIZING MEAT
Boiled meat: Add a tablespoon of vinegar to the cooking water.

Tough meat or game: Make a marinade of equal parts cooking vinegar and heated bouillon. Marinate for 2 hours.

Steak: Simply rub in a mixture of cooking vinegar and oil. Allow to stand for 2 hours.

Chicken: To stew an old hen, soak it in vinegar for several hours before cooking. It will taste like a spring chicken.

## INSTANT WHITE SAUCE
Blend together 1 C. soft butter and 1 C. flour. Spread in an ice cube tray, chill well, cut into 16 cubes before storing in a plastic bag in the freezer. For medium-thick sauce, drop 1 cube into 1 C. of milk and heat slowly, stirring as it thickens.

## UNPLEASANT COOKING ODORS
While cooking vegetables that give off unpleasant odors, simmer a small pan of vinegar on top of the stove. Or, add vinegar to the cooking water. To remove the odor of fish from cooking and serving implements, rinse in vinegar water.

## DON'T LOSE THOSE VITAMINS
Put vegetables in water after the water boils - not before - to be sure to preserve all the vegetables' vitamins.

## CLEAN AND DEODORIZE YOUR CUTTING BOARD
Bleach it clean with lemon juice. Take away strong odors like onion with baking soda. Just rub in.

## KEEP THE COLOR IN BEETS
If you find that your beets tend to lose color when you boil them, add a little lemon juice.

## NO-SMELL CABBAGE
Two things to do to keep cabbage smell from filling the kitchen; don't overcook it (keep it crisp) and put half a lemon in the water when you boil it.

## A GREAT ENERGY SAVER
When you're near the end of the baking time, turn the oven off and keep the door closed. The heat will stay the same long enough to finish baking your cake or pie and you'll save all that energy.

## GRATING CHEESE
Chill the cheese before grating and it will take much less time.

## SPECIAL LOOKING PIES
Give a unique look to your pies by using pinking shears to cut the dough. Make a pinked lattice crust!

## REMOVING HAM RIND
Before placing ham in the roasting pan, slit rind lengthwise on the underside. The rind will peel away as the ham cooks, and can be easily removed.

## SLUGGISH CATSUP
Push a drinking straw to the bottom of the bottle and remove. This admits enough air to start the catsup flowing.

## UNMOLDING GELATIN
Rinse the mold pan in cold water and coat with salad oil. The oil will give the gelatin a nice luster and it will easily fall out of the mold.

## LEFTOVER SQUASH
Squash that is leftover can be improved by adding some maple syrup before reheated.

## NO-SPILL CUPCAKES
An ice cream scoop can be used to fill cupcake papers without spilling.

## SLICING CAKE OR TORTE
Use dental floss to slice evenly and cleanly through a cake or torte - simply stretch a length of the floss taut and press down through the cake.

## CANNING PEACHES
Don't bother to remove skins when canning or freezing peaches. They will taste better and be more nutritious with the skin on.

## ANGEL FOOD COOKIES
Stale angel food cake can be cut into ½-inch slices and shaped with cookie cutters to make delicious "cookies". Just toast in the oven for a few minutes.

## HOW TO CHOP GARLIC
Chop in a small amount of salt to prevent pieces from sticking to the knife or chopping board then pulverize with the tip or the knife.

## EXCESS FAT ON SOUPS OR STEWS
Remove fat from stews or soups by refrigerating and eliminating fat as it rises and hardens on the surface. Or add lettuce leaves to the pot - the fat will cling to them. Discard lettuce before serving.

## BROILED MEAT DRIPPINGS
Place a piece of bread under the rack on which you are broiling meat. Not only will this absorb the dripping fat, but it will reduce the chance of the fat catching on fire.

## FAKE SOUR CREAM
To cut down on calories, run cottage cheese through the blender. It can be flavored with chives, extracts, etc., and used in place of mayonnaise.

## BROWNED BUTTER
Browning brings out the flavor of the butter, so only half as much is needed for seasoning vegetables if it is browned before it is added.

## COOKING DRIED BEANS
When cooking dried beans, add salt after cooking; if salt is added at the start it will slow the cooking process.

## TASTY CARROTS
Adding sugar and horseradish to cooked carrots improves their flavor.

## CARROT MARINADE
Marinate carrot sticks in dill pickle juice.

## CLEAN CUKES
A ball of nylon net cleans and smooths cucumbers when making pickles.

## FRESH GARLIC
Peel garlic and store in a covered jar of vegetable oil. The garlic will stay fresh and the oil will be nicely flavored for salad dressings.

## LEFTOVER WAFFLES
Freeze waffles that are left; they can be reheated in the toaster.

## FLUFFY RICE
Rice will be fluffier and whiter if you add 1 tsp. of lemon juice to each quart of water.

## NUTRITIOUS RICE
Cook rice in liquid saved from cooking vegetables to add flavor and nutrition. A nutty taste can be achieved by adding wheat germ to the rice.

## PERFECT NOODLES
When cooking noodles, bring required amount of water to a boil, add noodles, turn heat off and allow to stand for 20 minutes. This prevents overboiling and the chore of stirring. Noodles won't stick to the pan with this method.

## EASY CROUTONS
Make delicious croutons for soup or salad by saving toast, cutting into cubes, and sauteeing in garlic butter.

## BAKED FISH
To keep fish from sticking to the pan, bake on a bed of chopped onion, celery and parsley. This also adds a nice flavor to the fish.

### NON-STICKING BACON
Roll a package of bacon into a tube before opening. This will loosen the slices and keep them from sticking together.

### TASTY HOT DOGS
Boil hot dogs in sweet pickle juice and a little water for a different tate.

### GOLDEN-BROWN CHICKEN
For golden-brown fried chicken, roll it in powdered milk instead of flour.

### DOUBLE BOILER HINT
Toss a few marbles in the bottom of a double boiler. When the water boils down, the noise will let you know!

### FLOUR PUFF
Keep a powder puff in your flour container to easily dust your rolling pin or pastry board.

### JAR LABELS
Attach canning labels to the lids instead of the sides of jelly jars, to prevent the chore of removing the labels when the contents are gone.

### DIFFERENT MEATBALLS
Try using crushed corn flakes or corn bread instead of bread crumbs in a meatball recipe or use onion-flavored potato chips.

## CLEAN-UP TIPS

### APPLIANCES
To rid yellowing from white appliances try this. Mix together ½ C. bleach, ¼ C. baking soda, and 4 C. warm water. Apply with a sponge and let set for 10 minutes. Rinse and dry thoroughly.

Instead of using commercial waxes, shine with rubbing alcohol.

For quick clean-ups, rub with equal parts water and household ammonia.

Or, try club soda. It cleans and polishes at the same time.

## BLENDER
Fill part way with hot water and add a drop of detergent. Cover and turn it on for a few seconds. Rinse and drain dry.

## BREADBOARDS
To rid cutting board of onion, garlic or fish smell, cut a lime or lemon in two and rub the surface with the cut side of the fruit.

Or, make a paste of baking soda and water and apply generously. Rinse.

## COPPER POTS
Fill a spray bottle with vinegar and add 3 T. of salt. Spray solution liberally on copper pot. Let set for awhile, then simply rub clean.

Dip lemon halves in salt and rub.

Or, rub with Worcestershire sauce or catsup. The tarnish will disappear.

Clean with toothpaste and rinse.

## BURNT AND SCORCHED PANS
Sprinkle burnt pans liberally with baking soda, adding just enough water to moisten. Let stand for several hours. You can generally lift the burned portions right out of the pan.

Stubborn stains on non-stick cookware can be removed by boiling 2 T. of baking soda, ½ C. vinegar, and 1 C. water for 10 minutes. Re-season with salad oil.

## CAST-IRON SKILLETS
Clean the outside of the pan with commercial oven cleaner. Let set for 2 hours and the accumulated black stains can be removed with vinegar and water.

## CAN OPENER
Loosen grime by brushing with an old toothbrush. To thoroughly clean blades, run a paper towel through the cutting process.

## ENAMELWARE CASSEROLE DISHES
Fill a dish that contains stuck food bits with boiling water and 2 T. of baking soda. Let it stand and wash out.

## DISHES
Save time and money by using the cheapest brand of dishwashing detergent available, but add a few tablespoons of vinegar to the dishwasher. The vinegar will cut the grease and leave your dishes sparkling clean.

Before washing fine china and crystal, place a towel on the bottom of the sink to act as a cushion.

To remove coffee or tea stains and cigarette burns from fine china, rub with a damp cloth dipped in baking soda.

## DISHWASHER
Run a cup of white vinegar through the entire cycle in an empty dishwasher to remove all soap film.

## CLOGGED DRAINS
When a drain is clogged with grease, pour a cup of salt and a cup of baking soda into the drain followed by a kettle of boiling water. The grease will usually dissolve immediately and open the drain.

Coffee grounds are a no-no. They do a nice job of clogging, especially if they get mixed with grease.

## GARBAGE DISPOSAL
Grind a half lemon or orange rind in the disposal to remove any unpleasant odor.

## OVEN
Following a spill, sprinkle with salt immediately. When oven is cool, brush off burnt food and wipe with a damp sponge.

Sprinkle bottom of oven with automatic dishwasher soap and cover with wet paper towels. Let stand for a few hours.

A quick way to clean oven parts is to place a bath towel in the bathtub and pile all removable parts from the oven onto it. Draw enough hot water to just cover the parts and sprinkle a cup of dishwasher soap over it. While you are cleaning the inside of the oven, the rest will be cleaning itself.

An inexpensive oven cleaner. Set oven on warm for about 20 minutes, then turn off. Place a small dish of full strength ammonia on the top shelf. Put large pan of boiling water on the bottom shelf and let it set overnight. In the morning, open oven and let it air a while before washing off with soap and water. Even the hard baked-on grease will wash off easily.

## PLASTIC CUPS, DISHES AND CONTAINERS
Coffee or tea stains can be scoured with baking soda.

Or, fill the stained cup with hot water and drop in a few denture cleanser tablets. Let soak for 1 hour.

To rid foul odors from plastic containers, place crumbled-up newspaper (black and white only) into the container. Cover tightly and leave overnight.

## REFRIGERATOR
To help eliminate odors fill a small bowl with charcoal (the kind used for potted plants) and place it on a shelf in the refrigerator. It absorbs ordors rapidly.

An open box of baking soda will absorb food odors for at least a month or two.

A little vanilla poured on a piece of cotton and place in the refrigerator will eliminate odors.

To prevent mildew from forming, wipe with vinegar. The acid effectively kills the mildew fungus.

Use a glycerine-soaked cloth to wipe sides and shelves. Future spills wipe up easily. Add after the freezer has been defrosted, coat the inside coils with glycerine. The next time you defrost, the ice will loosen quickly and drop off in sheets.

Wash inside and out with a mixture of 3 T. of baking soda in a quart of warm water.

## SINKS
For a sparkling white sink, place paper towels across the bottom of your sink and saturate with household bleach. Let set for ½ hour or so.

Rub stainless steel sinks with lighter fluid if rust marks appear. After the rust disappears, wipe with your regular kitchen cleanser.

Use a cloth dampened with rubbing alcohol to remove water spots from stainless steel.

Spots on stainless steel can also be removed with white vinegar.

Club soda will shine up stainless steel sinks in a jiffy.

## SPONGES
Wash in your dishwasher or soak overnight in salt water or baking soda added to water.

## THERMOS BOTTLE
Fill the bottle with warm water, add 1 tsp. of baking soda and allow to soak.

## TIN PIE PANS
Remove rust by dipping a raw potato in cleaning powder and scouring.

## FINGERPRINTS OFF THE KITCHEN DOOR AND WALLS
Take away fingerprints and grime with a solution of half water and half ammonia. Put it in a spray bottle from one of these expensive cleaning products, you'll never have to buy them again.

## FORMICA TOPS
Polish them to a sparkle with club soda.

# KEEPING FOODS FRESH AND FOOD STORAGE

## CELERY AND LETTUCE
Store in refrigerator in paper bags instead of plastic. Leave the outside leaves and stalks on until ready to use.

## ONIONS
Once an onion has been cut in half, rub the leftover side with butter and it will keep fresh longer.

## CHEESE
Wrap cheese in a vinegar-dampened cloth to keep it from drying out.

## MILK
Milk at room temperature may spoil cold milk, so don't pour milk back into the carton.

## BROWN SUGAR
Wrap in a plastic bag and store in refrigerator in a coffee can with a snap-on lid.

## COCOA
Store cocoa in a glass jar in a dry and cool place.

## CAKES
Putting half an apple in the cake box will keep cake moist.

## ICE CREAM
Ice cream that has been opened and returned to the freezer sometimes forms a waxlike film on the top. To prevent this, after part of the ice cream has been removed press a piece of waxed paper against the surface and reseal the carton.

## LEMONS
Store whole lemons in a tightly sealed jar of water in the refrigerator. They will yield much more juice than when first purchased.

## LIMES
Store limes, wrapped in tissue paper, on lower shelf of the refrigerator.

## SMOKED MEATS
Wrap ham or bacon in vinegar-soaked cloth, then in waxed paper to preserve freshness.

## STRAWBERRIES
Keep in a colander in the refrigerator. Wash just before serving.

# NOTES

# NOTES

# NOTES

# NOTES

# NOTES